THE COMMONWEALTH AND INTERNATIONAL LIBRARY

Joint Chairmen of the Honorary Editorial Advisory Board

SIR ROBERT ROBINSON, O.M., F.R.S., LONDON
DEAN ATHELSTAN SPILHAUS, MINNESOTA
Publisher: ROBERT MAXWELL, M.C., M.P.

FOOD SCIENCE AND TECHNOLOGY
General Editor: PROFESSOR J. HAWTHORN

TECHNOLOGY OF CEREALS
with special reference to wheat

TECHNOLOGY OF CEREALS

with special reference to wheat

by

N. L. KENT, M.A., Ph.D. (Cantab.)

Research Association of British Flour-Millers, St. Albans, Herts.
Sometime Scholar of Emmanuel College, Cambridge

PERGAMON PRESS

OXFORD · LONDON · EDINBURGH · NEW YORK
TORONTO · PARIS · FRANKFURT

PERGAMON PRESS LTD.
Headington Hill Hall, Oxford
4 & 5 Fitzroy Square, London W.1

PERGAMON PRESS (SCOTLAND) LTD.
2 & 3 Teviot Place, Edinburgh 1

PERGAMON PRESS INC.
44–01 21st Street, Long Island City, New York 11101

PERGAMON OF CANADA LTD.
6 Adelaide Street East, Toronto, Ontario

PERGAMON PRESS S.A.R.L.
24 rue des Ecoles, Paris 5e

PERGAMON PRESS GmbH
Kaiserstrasse 75, Frankfurt-am-Main

First edition 1966

Library of Congress Catalog Card No. 65–26893

Printed in Great Britain by Watmoughs Limited, Idle, Bradford

(2384/66)

CONTENTS

PREFACE

THIS introduction to the technology of the principal cereals is intended, in the first place, for the use of students of Food Science. A nutritional approach has been chosen, and the effects of processing treatments on the nutritive value of the products has been emphasized. Throughout, both the merits and the limitations of individual cereals as sources of food products have been considered in a comparative way.

I am greatly indebted to Dr. T. Moran, C.B.E., Director of Research, for his encouragement and advice, and to all my senior colleagues in the Research Association of British Flour-Millers for their considerable help in the writing of this book. My thanks are also due to Miss R. Bennett of the British Baking Industries Research Association and Mr. R. Butler of the Ryvita Co. Ltd. who have read individual chapters and offered valuable criticism, and particularly to Professor J. A. Johnson of Kansas State University and Professor J. Hawthorn of The University of Strathclyde, Glasgow, who have read and criticized the whole of the text.

I wish to thank the firms which supplied pictures or data, viz. Henry Simon Ltd., Thos. Robinson & Son Ltd., Quaker Oats Ltd., Kellogg Co. of Great Britain Ltd., and also the authors, editors and publishers who have allowed reproduction of illustrations, including the Controller of H.M.S.O. for permission to reproduce Crown copyright material (Figs. 38 and 39, and data in Tables 1, 19, 23, 45 and 61).

Research Association of British Flour-Millers,   N. L. KENT
Cereals Research Station,
St. Albans, Herts.
July 1964

UNITS AND ABBREVIATIONS

THE following list is included for easy reference by the reader.

UNITS

ac	acre	MeV	megaelectronvolts
atm	atmosphere(s)	mg	milligramme (0·001 g)
Bé	Baumé		
Btu	British thermal units	min	minute(s)
bu	bushels	mm	millimetre
c	curie	μ	micro, micron (0·001 mm)
°C	degree Centigrade		
cwt	hundredweight (112 lb)	μg	microgramme (0·001 mg)
°F	degree Fahrenheit	μμc	micromicrocurie
ft	foot, feet	oz	ounce
g	gramme	ppm	parts per million
gal	gallon	sk	sack (280 lb of flour)
hr	hour	ton	long ton (2240 lb). The long ton is used throughout this book
in.	inch		
kg	kilogramme		
lb	pound		

ABBREVIATIONS

an.	annum	FFA	free fatty acid
Bk.	break	HRS	Hard Red Spring wheat
B.P.	boiling point		
Ch.	chapter	HRW	Hard Red Winter wheat
d.m.b.	dry matter basis		
E	extensibility (of dough)	m.c.	moisture content (wet basis)

mol. molecular

NRRL Northern Regional Research and Development Laboratory, Peoria, Ill., U.S.A.; a laboratory of the Northern Utilization Research and Development Division, Agricultural Research Service, U.S. Dept. of Agriculture

p., pp. page, pages

pt. parts

R resistance of dough

r.h. relative humidity

rev/min revolutions per minute

S spring (resistance) of dough

sp. species

sp. gr. specific gravity

SRW Soft Red Winter wheat

U.K. United Kingdom

U.S.A. United States of America

vac. vacuum

w. wire bolting cloth

wt. weight

CHAPTER I

CEREALS OF THE WORLD: CROPS

CEREALS

Cereals are the fruits of cultivated grasses, members of the family Gramineae. The principal cereal crops are wheat, maize, rice, barley, oats, rye, sorghum and millet. This book will deal mainly with the first six of these.

WORLD CROPS

The world production of the six cereals (wheat, maize, barley, oats, rye, rice) averaged about 656 million tons per annum outside China over the five-year period 1958–9 to 1962–3.[1] Throughout this book, tonnages refer to the long ton of 2240 lb. Production in China is not known accurately, but for wheat and rice (milled equivalent[2]) alone it was estimated at 107 million tons in 1958–9. The total production would thus provide approximately 580 lb of cereal grain per head per annum if shared equally among the entire world population. The average human consumption of cereals is considerably less than this, and a variable proportion of the six cereals is used for purposes other than human food, mainly animal feed, industrial processing, and seed.

The area occupied by the same six cereals, outside China, over the five-year period 1958–9 to 1962–3 averaged about 1150

[1] Throughout this book, the data for acreages, production, yields, crop movements and utilization of wheat, maize, barley, oats, rye and rice have been extracted or derived from the Commonwealth Economic Committee's publications *Grain Crops* (annually) and *Grain Bulletin* (monthly) by permission of the Controller of H.M.S.O.

[2] Production of rice (milled equivalent): rice is harvested as paddy, but rice production statistics are generally given in terms of the quantity of *milled rice* produced. 100 lb of paddy yield 63–73 lb of milled rice. See pp. 237–9.

million acres per annum. The 1960–1 area in China was estimated at about 145 million acres. Altogether, the acreage laid down to cereals represents approximately 3·7% of the entire land surface of the world. The annual acreage and production, outside China, of the six cereals since 1955–6 is shown diagrammatically in Fig. 1.

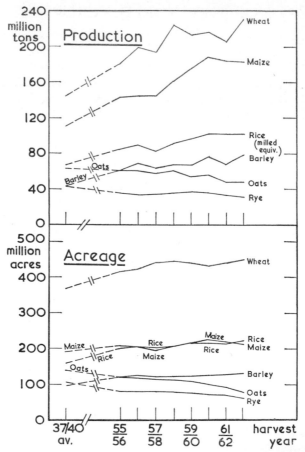

FIG. 1. Annual world production and acreage, outside China, of the six principal cereals since 1955–6. (Constructed from data in *Grain Crops* and *Grain Bulletin*.)

The percentage distribution of the world cereal acreage and production (outside China) among the six cereals, and the world average yields over the period 1958–9 to 1961–2 are shown in Table 1.

TABLE 1

WORLD CEREAL ACREAGE, PRODUCTION AND YIELD FOR THE PERIOD 1958–9 TO 1961–2 (EXCLUDING CHINA)

Cereal	Percentage of total acreage	Percentage of total production	Average yield* (cwt/ac)
Wheat	38	33	9·8
Maize	17	27	18·5
Barley	11	11	11·0
Oats	9	8	10·7
Rye	6	6	9·8
Rice	19	15†	9·2†

* Total world production divided by total world acreage.

† Milled equivalent: with a milling yield of 65%, the yield of paddy would be about 14 cwt/ac.

Data derived from *Grain Crops*, 1963, No. 9.

Proportion of acreage contributed by each cereal is similar to the proportion of production, except for maize. This is because the yields of wheat, barley, oats, rye and milled rice do not vary greatly among themselves, whereas the yield of maize is nearly double that of the other cereals. Taking all the cereals together, the average yield for the whole world outside China increased progressively from 9 cwt/ac in 1937–40 to 11·5 cwt/ac in 1961–2. Over the same period, the yield of maize increased by 8 cwt/ac (from 11·6 to 19·7 cwt/ac), largely through the use of hybrid maize (cf. p. 7), whereas that of other cereals increased by only 1–2 cwt/ac on average.

Wheat

The cultivation of wheat (*Triticum* sp.) reaches far back into history, and the crop was predominant in antiquity as a source of human food. It was cultivated particularly in Persia, Egypt, Greece and Europe. Numerous samples of ancient wheat have

been unearthed in archaeological investigations; the grains are
always carbonized, although in some the anatomical structure is
well preserved.

The characteristics of world wheat types and details of wheat
grading are given in Ch. 4.

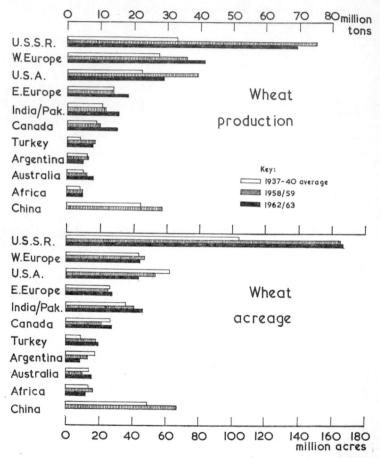

Fig. 2. Production and acreage of wheat in the main producing countries and areas of the world pre-war and in 1958–9 and 1962–3. (Constructed from data in *Grain Crops* and *Grain Bulletin*.)

Acreage, production and yield. The acreage and production of wheat in the principal countries and areas of the world, for three selected seasons, are shown diagrammatically in Fig. 2. Over the period 1958–9 to 1962–3 the Soviet Union produced about 31% of the world crop (outside China), western Europe 17%, the U.S.A. 15%, eastern Europe and India/Pakistan each 6–7%.

The yield per acre varies considerably among producing countries, and is related to the water supply and the intensity of cultivation. In Scandinavia and the Benelux countries, yields of over 30 cwt/ac, in other north-western European countries and in Japan of 20–28 cwt/ac, are obtained, whereas in more primitive agricultural communities, and in countries with less favourable climatic conditions, yields are still around 10 cwt/ac, and as low as 6–7 cwt/ac in India/Pakistan. The improvements in wheat yield since the war have occurred largely in the countries that are more advanced economically.

Wheat yield also depends on the type of wheat sown: winter wheat, with a longer growing period than spring wheat, is normally higher yielding than spring wheat. Durum wheat, which is grown in drier areas, is particularly low yielding.

The present yield of wheat in the U.K., 32 cwt/ac (1962–3), is approaching double the pre-war figure (18 cwt/ac). In the U.S.A., the recent increase in yield (7 cwt/ac pre-war; 13·4 cwt/ac in 1962–3) has been due to more widespread use of fertilizers, the growing of higher-yielding varieties, and progressive improvements in cultural methods.

Crop movements: exports. About 18% of the entire world wheat crop (outside China) was exported from the producing country to other countries over the period 1959–62. Of the total exports of wheat and wheat flour, about 38% was provided by the U.S.A., 22% by Canada, 13% by the Soviet Union, 11% by Australia, and 6% by Argentina.

About nine-tenths of the wheat exports are in the form of unmilled grain, the remainder being flour. The major exporters of wheat flour, from 1959 to 1962, were the U.S.A. (35% of the total), Western Germany (15%), Canada (18%), Australia

(13%) and France (8%). American flour goes to a large number of countries; two-fifths of Canadian flour comes to the U.K.; France has big markets for flour in her former African territories.

Crop movements: imports. Western European imports accounted for over one-third of the world movement of wheat over the period 1958–61, those of India and Pakistan for about one-seventh. Other large importers of wheat were Japan, Brazil and Poland.

Wheat flour is imported by the U.K., Egypt, Netherlands, Ceylon and Malaya: these countries absorb about 40% of the total world trade in wheat flour. Imports of flour decrease sharply when a domestic flour-milling industry is established: this happened recently in the Philippines.

Utilization. In the U.S.A., 82% of the wheat consumed domestically between 1957–8 and 1960–1 was used for human food, 7% for animal feed, and 11% for seed. In 1961–2 usage for animal feed increased to 11%, while usage for seed fell to 9%. Industrial usage was about 18,000 tons in 1955–6, falling to 2000 tons per annum during 1959–62, or less than 0·02% of the total domestic consumption. Domestic utilization of wheat in the U.K. during the period 1957–8 to 1961–2 (comprising 34–39% of home-grown new crop plus 61–66% of imports) was 73–77% for human food, 20–25% for animal feed, 2% for seed, and less than 0·1% for industrial purposes.

Maize

Maize (corn, in the U.S.A.) (*Zea mays*) is cultivated in regions that experience periods of at least 90 days of frost-free conditions.

The rainfall where it is grown may vary from 10 to 200 in./an. but suitable types are available for these varying conditions.

The crop originated in America, where it is widely grown, and it has been taken to Africa, India, Australia, and the warmer parts of Europe.

The principal types of maize are: dent and flint (the varieties of commerce: cf. p. 31), pod, pop, soft or flour, sweet and waxy (cf. p. 40).

Hybrid maize. The work of Shull and East in the U.S.A., early this century, led to a revolution in maize breeding that has produced remarkable results. Hybrid lines of maize have been developed that give 15–20%, sometimes up to 50%, higher yields than those of inbred lines, the use of which has been almost superseded in the U.S.A. The farmer generally obtains fresh hybrid seed each year from growers who specialize in its production. The grower chooses an isolated field on which he grows two inbred lines of maize in alternate strips, one row of male parent plants to about four rows of female parent plants. At the appropriate time, the female parent plants are de-tasselled, and are subsequently fertilized by pollen from the adjacent male parent plants. Seed is later collected from the female parent plants only.

Maize grains may be white, yellow or reddish in colour, and there are definite colour preferences in some of the consumer countries. White is preferred in western Europe, but yellow is preferred for poultry feeds; the reddish type is favoured in Japan.

Grading. In the U.S.A., maize is classified as White, Yellow, or Mixed, and is further qualified as Flint, or Flint and Dent (cf. p. 31). Five numerical grades are recognized, with characteristics as shown in Table 2.

TABLE 2
U.S. GRADES OF MAIZE*

Grade No.	Minimum test wt. (lb/bu)	Moisture (%)	Broken† maize or foreign material (%)	Damaged kernels	
				Total (%)	Heat damaged (%)
1	56	14·0	2·0	3·0	0·1
2	54	15·5	3·0	5·0	0·2
3	52	17·5	4·0	7·0	0·5
4	49	20·0	5·0	10·0	1·0
5	46	23·0	7·0	15·0	3·0

* From U.S. Dept. Agric. (1961).
† Passing through a 12/64 in. round hole sieve.

Maize is described as Sample Grade if it does not meet the requirements of grades 1–5, or contains stones, or is musty, sour or heating, or has any commercially objectionable foreign odour, or is of distinctly low quality.

Acreage, production and yield. During the immediate pre-war period, about 45% of the world acreage of maize grown outside China was in the U.S.A., no other single country having more than about 5% of the total acreage. Other principal producing countries were Brazil, the Soviet Union, Roumania, Argentina, Mexico, India/Pakistan, Yugoslavia and Indonesia. At that time, the U.S.A. produced about 53% of the world crop, as yields in the U.S.A. were somewhat higher than those elsewhere. By 1956–7, however, the U.S.A.'s share of the world acreage (outside China) had shrunk to 31%, and in 1961–2 fell further to 26%, while that of the Soviet Union had increased to 10% by 1958–9, and to 15% by 1961–2, the contribution of other countries showing less variation. Yet the U.S.A. continued to produce half of the world's total maize tonnage over the whole period because of the development, in recent years, of high-yielding strains of hybrid maize, which are now grown there almost exclusively (cf. above, p. 7).

Yields in the U.S.A. increased from 11 cwt/ac in 1934–8 to 27 cwt/ac in 1960–1; yields in excess of 25 cwt/ac are now obtained also in Austria and Italy, and yields of 20–25 cwt/ac in Southern Rhodesia and France, but only 12–19 cwt/ac in Argentina, Roumania, Yugoslavia and Egypt, and less than 10 cwt/ac in India, Mexico and Indonesia.

Crop movements. The proportion of the world crop, excluding that grown in China and the Soviet Union, that entered into world trade was about 5% in 1937–40, gradually rising to 9% in 1962.

In 1937–40 the U.S.A. was exporting about 1·5 million tons of maize per annum, but by 1962 the figure had increased to 10·6 million tons. The other principal exporting countries in 1962 were Argentina (2·8 million tons, a considerable reduction on pre-war figures) and the Republic of South Africa (2 million tons), contributing to the world total of about 17·5 million tons of exports apart from those from the Soviet Union and China.

The U.K. is the biggest importer of maize. The annual figure was about 3 million tons in 1937–40, falling to half this in 1955–7, but reaching 3 million tons again in 1959–61, and 4·5 million tons in 1962. In 1962, Italy took 2·7 million tons, Japan 2·3 million tons, and Netherlands and all Germany each about 1·5 million tons.

Utilization. See Ch. 14, p. 244.

Barley

Barley is grown in temperate climates as a spring crop, and has geographic distribution generally similar to that of wheat. Barley grows well on well-drained soils, which need not be so fertile as those required by wheat. Some winter barleys are grown.

Classification. The genus *Hordeum*, which includes the barleys, was divided by Linnaeus into six species, viz. wide- and narrow-eared six-row barley, wide- and narrow-eared two-row barley, and two species of naked barley. Subsequently, taxonomists have regarded these as sub-groups of one chief species, named *H. sativum* or *H. vulgare*. For present purposes, barleys may be considered as belonging to three main types: (a) hulled, six-row; (b) hulled, two-row; (c) hull-less.

In the hulled types, the hull or husk is adherent to the kernel after threshing, whereas in the hull-less types it is loose and easily removed during threshing. Two- or six-row refers to the arrangement of the grains on the spike.

Hulled barley. The barley grown in Britain, north-western Europe and Australia is mostly of the hulled two-row type, which seems well suited to the light and medium soils and the prevailing moderate temperature and rainfall. Hulled six-row barley is also grown as a winter crop in western Europe. Both two- and six-row hulled barleys have been introduced into the U.S.A. in comparatively recent times: of these, the six-row type is more adaptable to varied environmental conditions, outyields the two-row type, and is preferred both by maltsters and farmers.

The six-row type is also predominant in India and the Middle East.

Hull-less barley. Hull-less or naked types of barley are culti-vated extensively in south-east Asia. The yield of grain is lower than that of the hulled types, and the spikelets have a tendency to shed grain when ripe, thus further reducing the yield. Hull-less types have weaker straw and are more liable to lodge than the hulled types. The absence of hull makes them unsuitable for malting (cf. p. 199), but they are useful for food, having a higher digestibility (94%) than the hulled types (83%). In 1960–1, approximately equal quantities of hulled and hull-less (naked) barleys were produced and used for food in Japan.

U.S. grades of barley. White-hulled barley grown east of the Rocky Mountains is divided into three subclasses as follows:

1. *Malting barley:* six-rowed barley, which has 90% or more of the kernels with white aleurone layer, and which is not semi-steely in the mass;
2. *Blue malting barley:* six-rowed barley, which has 90% or more of the kernels with blue aleurone layer, but otherwise as malting barley;
3. *Barley:* barley that does not meet the requirements of the other subclasses.

Each subclass is graded into numerical grades. The grade requirements for malting and blue malting barley are shown in Table 3.

TABLE 3
U.S. GRADING OF MALTING AND BLUE MALTING BARLEY*

| Grade | Minimum limits of | | Maximum limits of | | | | | |
	Test wt. per bu (lb)	Sound barley (%)	Damaged kernels (%)	Foreign material (%)	Skinned and broken kernels (%)	Thin barley (%)	Black barley (%)	Other grains (%)
1	47	97	2	1	4	7	0·5	2
2	45	94	3	2	6	10	1	3
3	43	90	4	3	8	15	2	5

* From U.S. Dept. Agric. (1961).

Acreage, production and yield. The world acreage, outside China, laid down to barley, increased from 98 million in 1937–40 to 127 million in 1956–7, and since then has fluctuated between 120 and 130 million, with production between 63 and 76 million tons per annum. Since 1961–2, the Soviet Union has contributed about one-quarter of the world's acreage but only one-fifth of the world's production, the U.S.A. one-tenth and one-eighth, respectively. France and the U.K. have each contributed 7–8% of the total production. Other principal producing countries include Turkey, India, Canada, Western Germany, Japan and Denmark.

The highest yields are obtained in the intensely cultivated areas: yields of 28 cwt/ac or more have been obtained in Denmark in most years since 1956, and reached 31 cwt/ac in 1962–3. In that season, yields of 29 cwt/ac were obtained in the U.K., 26 cwt/ac in Western Germany, and yields in excess of 20 cwt/ac in France, Sweden and Japan. Appreciably lower yields are obtained in North America and Australia, and only 6–7 cwt/ac in areas of the Middle East and North Africa.

Crop movement. The proportion of the world production of barley (outside China) that moved in world commerce rose from 5% in 1937–40 to 8% in 1962. The principal exporting countries in the immediate pre-war years were Canada, the Soviet Union, Argentina and the U.S.A., each exporting 11–15% of the total. In recent years, exports from Argentina have dwindled, whilst the U.S.A. and France have become the largest exporters. Australia, Canada and the Soviet Union each provided about 7% of the total exports in 1962.

European countries were the principal importers until 1961, when Chinese imports of over 1 million tons exceeded those of the U.K. (0·97 million tons) and Western Germany (0·96 million tons). In 1962, Chinese imports fell to 0·3 million tons, while those of Western Germany rose to 2 million tons.

Utilization. See Ch. 12, p. 199.

Oats

The oat crop (*Avena* sp.) is widely cultivated in temperate regions; it is more successful than wheat or barley in wet climates, although it does not stand cold so well. A small proportion of the crop is milled for human food (cf. Ch.13). Most of it, however, is used for animal feeding, although increased mechanization on farms has reduced the quantity of oats required for feeding horses. In the U.K., oats are grown extensively in Scotland and the north of England, where better quality crops are obtained than in the south.

U.S. grading of oats. In the U.S.A., "Oats" is defined as grain consisting of 50% or more of cultivated oats (*A. sativa* and/or *A. byzantina*), containing not more than 25% of wild oats and other grains for which standards have been established. Oats are classed as White, Red, Grey, Black or Mixed; each class is graded into four numerical grades and sample grade, as shown in Table 4.

TABLE 4
U.S. GRADES OF OATS*

	Minimum limits of		Maximum limits of		
Grade	Test wt. per bu (lb)	Sound cultivated oats (%)	Heat damaged kernels (%)	Foreign material (%)	Wild oats (%)
1	34	97	0·1	2·0	2·0
2	32	94	0·3	3·0	3·0
3	30	90	1·0	4·0	5·0
4	27	80	3·0	5·0	10·0

* From U.S. Dept. Agric. (1961).

Acreage, production and yield. The annual world acreage of the oat crop, outside China, steadily declined from 139 million for the 1937–40 average to 80 million in 1962–3, i.e. by 42%. The decline affected all the major producing countries, and particularly the Soviet Union, which in 1962–3 contributed only 21% to the total world acreage as compared with 35% pre-war and 31–34% during the period 1958–9 to 1961–2. Other major contributors to the world acreage during the period 1958-9 to

1962–3 were the U.S.A. 26–29%, Canada 8–13%, Poland 4–5% and France 3–4%.

The decline in production over the same period has been only 23%, viz. from 63 to 48 million tons annually, because of the steady increase in average yield per acre. Production has declined more steeply in the Soviet Union than in the other major producing countries on account of acreage restriction.

Pre-war yields of 19–21 cwt/ac in Denmark and Netherlands have gradually increased, yields of 29–30 cwt/ac being obtained in 1962–3. The increase in yield over the same period in the U.K., Belgium and Western Germany has been from 16 to 21 cwt/ac. In some countries, however, yields are much lower, e.g. only 6–9 cwt/ac in Australia and Argentina. The average yield in the Soviet Union (calculated from data for estimated acreage and production) has remained almost constant at about 7 cwt/ac since 1937–40.

Crop movement. The bulk of the oat crop is consumed on the farm where it is produced: only 1–3% of the total crop enters world commerce. Exports of oats have, however, increased markedly since 1937–40, despite falling production; the quantity exported from the principal exporting countries increased from 0·75 million tons in 1937–40 to a peak of 1·7 million tons in 1958, subsequently declining to 1·3 million tons in 1962. Exports from individual countries have fluctuated considerably over this period: in various years, Argentina, the U.S.A., Canada or Australia have been the biggest exporters. These four countries have, between them, contributed three-quarters or more of the total exports since 1955.

European countries are the biggest importers of oats, accounting for 85–95% of total imports in 1960–2. Western Germany, Netherlands and Switzerland were the individual countries taking the largest quantities.

Utilization. See Ch. 13, p. 211.

Rye

On good soil, rye (*Secale cereale*) is a less profitable crop than wheat, but on light acid soil it gives a more satisfactory yield. The rye crop nevertheless benefits from manurial treatment. It is more resistant than wheat to pests and diseases, and can better withstand cold. Because it has these characteristics, rye tends to be grown on land just outside the belt which gives the most satisfactory return with the wheat crop, such as areas of northern and eastern Europe that have a temperate climate.

The native rye grown in Britain until about 20 years ago was long-strawed, gave low yields and had high (10–13%) protein content. Since then, new varieties, e.g. King II, Dominant, Petkus and Pearl, have been introduced, particularly from Sweden and Germany; they are short-strawed and suitable for combine-harvesting, equal to wheat in yielding ability on good land, and low (7–8%) in protein content. Tetra Petkus is a tetraploid variety (i.e. with double the normal number of chromosomes) produced recently in Germany by treating Petkus rye with colchicine. The grain yield is reported to be 115% of that of Petkus, and the variety is credited with superior winter hardiness, test weight (bushel weight; see p. 70) and agronomic characteristics. The open or cross pollination of rye increases the difficulty of keeping the strain pure, and therefore new seed must be raised in isolation for this purpose.

U.S. grades of rye. Only one class of rye is recognized: it is defined as any crop which before removal of dockage consists of 50% or more of rye and not more than 10% of other grains for which standards have been established under provisions of the U.S. Grain Standards Act. The grades of rye recognized in the U.S.A., and their limiting characteristics, are shown in Table 5.

For all grades, the maximum permitted content of ergot (cf. p. 232) is 0·3%.

Acreage, production and yield. The annual world acreage of rye, 105 million in 1937–40, fell to 82–79 million during the period 1955–9, and to 63 million in 1962–3. The decline, as with oats, has been greatest in the Soviet Union, where the place of these two cereals has been taken by increased acreages of wheat, maize and barley. Over this period, the Soviet Union has accounted

TABLE 5

U.S. GRADES OF RYE*

Grade No.	Minimum test wt. (lb/bu)	Maximum limits for			
		Damaged kernels (rye and other grains)		Foreign material	
		Total (%)	Heat damaged (%)	Total (%)	Other than wheat (%)
1	56	2	0·1	3	1
2	54	4	0·2	6	2
3	52	7	0·5	10	4
4	49	15	3·0	10	6

* From U.S. Dept. Agric. (1961).

for 55–60% of the total world rye acreage, the only other major producing country being Poland (15–18% of the total acreage).

World acreage decline of 32% between 1937–40 and 1961–2 is matched by world production decline of 19% over the same period, the major part of the production decline occurring before 1955–6. Since then, production has been fairly steady at 34–37 million tons per annum. During the period 1956–7 to 1961–2, the Soviet Union's share of the world production increased from 41 to 47% and that of Poland from 19 to 24%; the contribution from Germany (West and East) decreased from 18 to 11%.

In the U.K., rye is now grown mostly in East Anglia and Yorkshire as a winter crop. The total acreage is about 19,000, and the national production about 0·5% of that of wheat.

In the U.S.A., rye is grown chiefly in the upper north central States, principally in North Dakota, where emigrants from rye-growing countries of eastern Europe have settled. Between 1957–8 and 1962–3, the acreage of the U.S. rye crop (1·5–2·0 million per annum) has been 3–4·5% of the U.S. acreage harvested for wheat; the U.S. rye crop (0·6–1·0 million tons per annum) is 2–3% of the world rye crop.

Yields exceeded 23 cwt/ac in 1960–1 in the western European countries practising intensive cultivation: Netherlands, Belgium, Denmark and Western Germany; but in the three principal producing countries yields were only 12 (Poland), 10 (U.S.A.) and 8 cwt/ac (Soviet Union).

Crop movement. The proportion of the world's rye crop moving in international commerce increased from 2% in 1937–40 to 6% in 1962. The Soviet Union is easily the biggest exporter, her share of the total exports steadily increasing from 31% in 1956 to 62% in 1962. Small quantities were exported pre-war by Poland and Roumania, and since the war by Western Germany and the U.S.A. The only other major sources are Argentina and Canada.

European countries are the major importers of rye, the largest quantities going to Germany (Western and Eastern), Poland and Netherlands. Western Germany both exports and imports rye. Imports come mainly from Canada, the U.S.A. and the Soviet Union; the rye grown in Western Germany in 1961 was exported mainly to Italy and Netherlands.

Utilization. See Ch.14, p. 227.

Rice

The rice crop (*Oryza* spp.) is grown in the tropics where rain and sunshine are abundant. Although typically a cereal of the swamp, rice can be grown either on dry land or under water. The common practice of flooding the paddy fields has been adopted as a means of irrigation and also as a means of controlling weeds. In much of Asia and Africa, however, rice is grown on hilly land without irrigation. In some Asian countries, where irrigation is practised, two crops of rice are grown per year. The main crop is grown in the wet season, the subsidiary crop in the dry season (with irrigation). Yields are lower in the main crop than in the subsidiary because of lack of sunshine.

There are varieties of rice adapted to a wide range of environmental conditions: it can be grown in hot wet climates, but equally in the foothills of the Alps, up to 4000 ft in the Andes of Peru, 6000 ft in the Philippines, and 10,000 ft in India. This wide adaptability of the rice plant is the explanation of its importance as a food crop.

In the U.S.A. rice is grown in two well-defined areas: Texas–Louisiana–Arkansas, and California. In these States the requirements of the rice crop are found—level land with an impervious subsoil, and abundant water for irrigation. Rice is a highly mechanized crop in the U.S.A., where planting, fertilizer treatment and weeding are all carried out on a large scale by means of aircraft. The crop is harvested by combine harvesters. In contrast, rice grown in the major producing countries, amounting to 90% of the world crop, is managed entirely without mechanization.

Acreage, production and yield. The world production of rice (paddy) is probably about equal to that of wheat. In 1958–9 both crops yielded about 250 million tons, but since that season the production of wheat in China has not been reported, and strict comparisons cannot be made (*N.B.*: the data in Table 1 and in Fig. 1 are for the world excluding China, and those for rice are given in terms of milled equivalent; see p. 239). This production is achieved on an acreage equivalent to about 60% of that of wheat, because the yield of rice per acre is so much the greater. Rice is the basic food for more than half the world's population, and provides up to 80% of the food intake in some countries.

The yield of foodstuffs energy produced per acre varies widely according to the kind of foodstuffs: more food, on an energy basis, is produced from one acre of rice than from one acre of any other cereal. With a milling yield of 65%, and a calorific value of 1600 kcal/lb for rice, Mickus (1959) has calculated that a yield of 3000 lb/ac produces almost 5 million kcal per acre per annum, which is 2–3 times that produced by wheat.

The total world acreage of rice in 1937–40 averaged 208 million per annum; India and Pakistan contributed 78 million, and the estimated acreage in China was 49 million. Cambodia, Laos and Vietnam together contributed 14 million acres, Burma 12·6 million, and Indonesia 10·4. After the war, the acreage in India and Pakistan increased to 107 million in 1959–62 and that in China to an estimated 82 million in 1956–7, declining to 75 million estimated in 1961–2. Other major producers are Thailand, Japan, the Philippines and Brazil.

China is by far the biggest producer of rice. Her estimated production in 1958–9 was 78 million tons (in terms of "milled equivalent"), compared with 92 million tons from the whole of the rest of the world, falling to 54 million tons in 1961–2 (rest of world: 102 million tons).

The yield of rice per acre varies widely according to the method of cultivation. In general, it is high in sub-tropical regions where the variety *japonica* is grown; contributory factors are the intensive cultivation practised in some areas in these regions, and the fact that *japonica* rice gives increased yield when heavily fertilized.

The highest yields are obtained in Australia, most of the crop being grown in New South Wales, where the average yield since 1958–9 has been 32–35 cwt/ac (milled equivalent). Yields (milled equivalent) of 32 cwt/ac in Spain, 30 cwt/ac in California, 29 cwt/ac in Italy, 27 cwt/ac in Japan, and 26 cwt/ac in Egypt, have been obtained since 1959–60. In the major producing countries of tropical south and south-east Asia yields are much lower: 8–10 cwt/ac in Ceylon, Indonesia, Burma, India and Pakistan; 7 cwt/ac in Thailand; and only 5–6 cwt/ac in the Philippines and Cambodia.

Crop movements. The proportions of the world crop moving in international commerce decreased from over 7% in 1937–9 to 3·4% in 1962. Burma and Thailand are the principal suppliers (45–56% of the annual total since 1956). Imports go to Indonesia, Malaya, India, Pakistan, Ceylon, Hong Kong and other Asian countries, with a relatively small proportion coming to European countries, principally Germany.

Utilization. See Ch. 14, p. 236.

Sorghum

Sorghum, Kafir corn, or Milo (*Sorghum vulgare*) is a coarse grass which bears loose panicles containing up to 2000 seeds per panicle. It is an important crop and the chief food grain in parts of Africa, Asia, India/Pakistan and China, where it forms a large part of the human diet.

The crop is grown in latitudes below 45° in all continents; in the U.S.A., it is grown in the Great Plains area, chiefly in Texas where it is the most important crop, and in Kansas. The most favourable mean temperature for the crop is 80°F and, although it does well in semi-arid conditions, it repays irrigation. The crop is not troubled by serious pests or diseases, and has the advantage that it can be sown late, in case other crops fail (Matz, 1959).

Acreage, production and yield. The world acreage was reported by Martin and Leonard (1949) as exceeding 80 million acres. The acreage in the U.S.A. averaged 9·8 million acres in 1948–57. Total U.S. production over this period was about 5 million tons, giving an average yield of 10·7 cwt/ac.

The U.S. sorghum crop has increased since 1956. The increase has been due to a change over from other crops (cotton, wheat, maize) because of U.S. Government agricultural programmes; to the availability of sorghum hybrids, first introduced in 1955, which give 20–40% higher yields than types previously available; and to a succession of drought years. The types now grown are suitable for combine harvesting.

Utilization. See Ch. 14, p. 254.

REFERENCES

Commonwealth Economic Committee, *Grain Crops*, H.M.S.O., London. An annual publication.

Commonwealth Economic Committee, *Grain Bulletin*, Intelligence Branch of C.E.C., Marlborough House, Pall Mall, London W.1. A monthly publication.

Martin, J. H. and Leonard, W. H. (1949), *Principles of Field Crop Production*, Macmillan, New York.

Matz, S. A. (1959) (Ed.), *The Chemistry and Technology of Cereals as Food and Feed*, Avi Publ. Co., Westport, Conn., U.S.A.

Mickus, R. R. (1959), Rice (*Oryza sativa*), *Cereal Sci. Today* 4: 138.

U.S. Department of Agriculture (1961), Official Grain Standards of the United States, Service and Regulatory Announcement AMS–177, issued June 1957, revised October 1961.

CHAPTER 2

CEREALS OF THE WORLD: STRUCTURE OF CEREAL GRAINS

FLOWER STRUCTURE

The cereals of commerce and industry are harvested, transported and stored in the form of grain. The grains develop from flowers or florets which, in wheat, barley, rye, oats and rice, comprise ovary, three stamens and two scale-like lodicules, all of which are enclosed in a pair of bracts or leaves called lemma and palea. The lemma covers the dorsal side of the floret (the side towards the rachis), whilst the palea covers the groove on the ventral side of the grain. The ovary and the ovule within it, after fertilization by pollen, develop into the grain.

In wheat and rye, the lemma and palea are loose and become free from the grain at threshing, forming the chaff. Wheat and rye are thus naked caryopses.

In hulled forms of barley, the lemma and palea fuse with the ovary as the grain develops, forming the hull or husk of the grain. Thus, barley is a covered or coated caryopsis. The lemma is awned and is larger than the palea.

In oats, the lemma and palea, which enclose the kernel when the grain is ripe, are tough and leathery, and although not fused to the kernel as in the case of barley, are nevertheless adherent to it, and do not come away during threshing: they form the hull or husk of oats which, like barley, is thus a covered caryopsis. The grains of rice, like those of most varieties of barley and oats, retain the hull or husk after threshing. The threshed rice is known as *paddy* or *rough rice*.

Maize

Male and female flowers are contained in separate inflorescences upon the same plant; the female inflorescence consists of a central rachis (the cob: cf. p. 253) bearing numerous rows of sessile spikelets, the whole being enclosed by a number of overlapping bracts which constitute the "husk". A tuft of thread-like silks emerges from the top of the spike: these are the stigmas of the flowers. Each spikelet in the female inflorescence comprises a pair of glumes and two florets, of which only one, the upper, is fertile. The grains of maize are naked caryopses.

ANATOMY OF CEREAL GRAINS

Size

The approximate range of sizes encountered among mature grains of the cereals, and the average weight per 1000 grains, are shown in Table 6.

TABLE 6

DIMENSIONS AND AVERAGE WEIGHT PER 1000 GRAINS OF THE CEREALS

Cereal	Dimensions		Average grain wt. (g per 1000)
	Length (mm)	Width (mm)	
Rye	4·5–10	1·5–3·5	21
Sorghum	3–5	2·5–4·5	23
Rice (paddy)	5–10	1·5–5	27
Oats	6–13	1–4·5	32
Wheat	5–8	2·5–4·5	37 { Manitoba: 27
Barley	8–14	1–4·5	37 { English: 48
Maize	8–17	5–15	285

Anatomical structure

The anatomical structure of all cereal grains is basically similar, differing from one cereal to another in detail only. The outline of the cereal grains is shown in Fig. 3.

Grains of wheat, rye and maize (naked caryopses) consist of fruit coat (pericarp) and seed. The seed is comprised of seed coat, germ and endosperm. Grains of oats, barley and rice (covered or

Naked caryopses

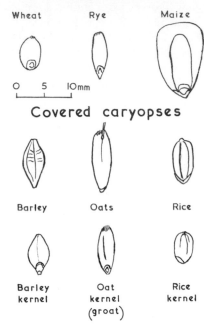

Covered caryopses

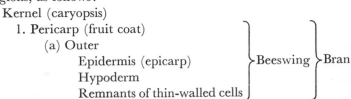

Fig. 3. Grains of the six principal cereals, showing comparative sizes and shapes. The kernels of the three husked grains (barley, oats, rice) are shown in the bottom row.

coated caryopses) have additionally, outside the fruit coat, the fused glumes (palea and lemma) which constitute the husk. Each of the main parts of the grain—pericarp, seed coat, germ and endosperm—is further subdivided into various layers, tissues or regions, as follows:

    Kernel (caryopsis)
        1. Pericarp (fruit coat)
            (a) Outer
                Epidermis (epicarp)   ⎫
                Hypoderm   ⎬ Beeswing ⎬ Bran
                Remnants of thin-walled cells ⎭

(b) Inner
  Intermediate cells
  Cross cells
  Tube cells
2. Seed   ⎫
  (a) Seed coat (testa) and pigment strand   ⎬ Bran
  (b) Nucellar layer (hyaline layer)   ⎭
  (c) Endosperm
    Aleurone layer
    Starchy endosperm
  (d) Germ (embryo)
    Scutellum (cotyledon)
    Embryonic axis
      Plumule, covered by coleoptile
      Primary root, covered by coleorhiza
      Secondary lateral roots
    Epiblast

Wheat

Wheat grains (see Figs. 3 and 4) are ovoid in shape, rounded at both ends. The germ is prominent at one end, a tuft of fine hairs (the beard) at the other. Along the ventral side of the grain there is an indentation or furrow (the "crease"), an infolding of the aleurone and all covering layers. At the bottom of the crease is a deeply pigmented vascular strand.

The presence of the crease complicates any milling process which aims at separating the endosperm from the enclosing layers. An abrasive pearling process that removes the bran and aleurone from the exterior of the grain does not reach the in-folded part of the bran in the crease. The flour-miller has solved the problem of the crease by using rollermills with fluted rolls which, in the initial stages of the process, open out the wheat grain to expose the endosperm in the form of irregular columns attached to the bran and, in later stages, remove most of the endosperm from the bran (see Fig. 5 and cf. p. 117).

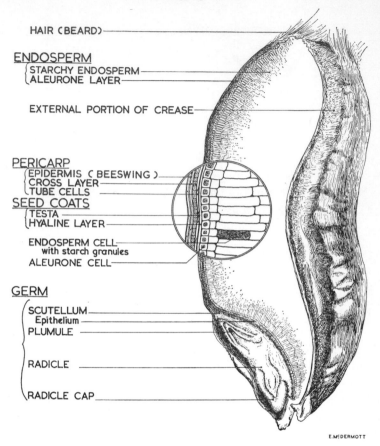

HAIR (BEARD)

ENDOSPERM
  STARCHY ENDOSPERM
  ALEURONE LAYER

EXTERNAL PORTION OF CREASE

PERICARP
  EPIDERMIS (BEESWING)
  CROSS LAYER
  TUBE CELLS
SEED COATS
  TESTA
  HYALINE LAYER

ENDOSPERM CELL
  with starch granules
ALEURONE CELL

GERM
  SCUTELLUM
  Epithelium
  PLUMULE

  RADICLE

  RADICLE CAP

E.McDERMOTT

Fig. 4. Longitudinal section of wheat grain through crease and germ.

Barley (Figs. 3 and 6)

The grain of barley is spindle-shaped, thicker in the centre, and tapering towards each end.

The hull of barley (in hulled types) protects the **grain** from predators, and serves a useful purpose in the malting and brewing processes (cf. p. 203 and Table 48). It amounts to about 13% of

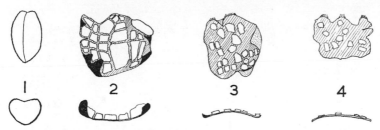

FIG. 5. Stages in opening out of the wheat grain and scraping of endosperm from bran by fluted rolls of the Break system. 1, Whole wheat grain; 2, I Bk. tails; 3, II Bk. tails; 4, III Bk. tails. *Upper row:* plan view (looking down onto inside of the bran in 2–4). *Lower side:* side view. In 2–4, endosperm particles adherent to bran are uncoloured; inner surface of bran, free of endosperm, is hatched; outer side of bran curling over is shown in solid black; beeswing, from which bran has broken away, is shown dotted.

the grain (by weight) on average, the proportions ranging from 7 to 25% according to type, variety, grain size, and latitude of cultivation. Winter barleys have more hull than spring types; six-row (12·5%) more than two-row (10·4%). The proportion of hull increases as the latitude where the barley is grown approaches the equator, e.g. 7–8% in Sweden, 8–9% in France, 13–14% in Tunisia. Large and heavy grains have less hull than small, lightweight grains.

Oats

The number of grains per spikelet is characteristic of the variety and of the environmental conditions of growth. Two per spikelet is typical, but some varieties (e.g. Potato oats) generally have only one, while others frequently have three. Diversity of grain size and of percentage kernel content increases with the number of grains per spikelet, and this is important in relation to the shelling process (cf. p. 219).

The oat grain (Figs. 3 and 7) is cylindrical in shape, blunt at the germ end, pointed at the beard end. The grain is comprised of the caryopsis or kernel, often called the "groat", and the enclosing hull or husk, formed from the lemma and palea. The

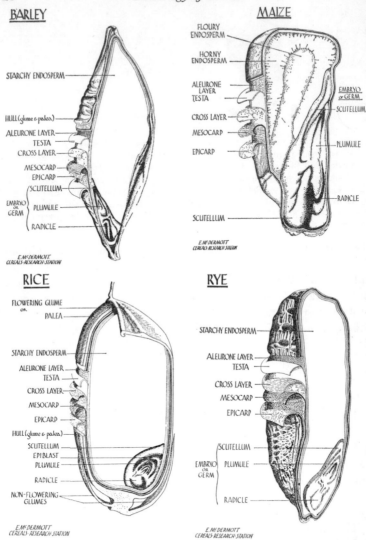

FIG. 6. Longitudinal sections through grains of barley, rye, maize, and rice. (The rice diagram has appeared in D. H. Grist, *Rice*, 3rd edition, 1959, and is reproduced by courtesy of the publishers, Longmans, London.)

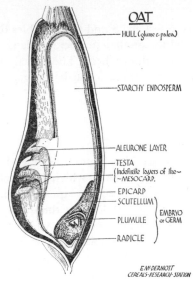

OAT

— HULL (glume & palea)

— STARCHY ENDOSPERM

— ALEURONE LAYER
— TESTA
{ Indefinite layers of the—
{ —MESOCARP.
— EPICARP
— SCUTELLUM }
                 } EMBRYO
— PLUMULE   } or GERM
— RADICLE   }

E. McDERMOTT
CEREALS-RESEARCH- STATION

FIG. 7. Longitudinal section through oat grain. (Reproduced from N. L. Kent, *Cereal Sci. Today* **2**: 83, 1957, by courtesy of the Editor.)

husk is a relatively low-priced by-product of the oatmeal milling process (cf. p. 224).

The maximum yield of oatmeal obtainable from a parcel of oats is closely related to the amount of kernel in the grain, and thus the percentage kernel content (wt. of kernel expressed as a percentage of the weight of kernel plus husk) is probably the most important quality characteristic of milling oats.

Kernel content ranges from 65 to 81% in cleaned British-grown oats, average 75%, higher kernel content (i.e. lower husk content) resulting in higher yield of oatmeal. The differences in kernel content between samples are due, in part, to variety and, in part, to environment: kernel content tends to be higher in winter-sown than in spring-sown varieties, in Scottish-grown than in English-grown samples, and in the small third grains than in the large first (main) grains of varieties with three grains per spikelet (Hutchinson *et al.*, 1952).

Rye

The grain of rye (see Figs. 3 and 6) resembles that of wheat in structure but is slightly smaller in average size and is longer in proportion to its width than wheat. The apical end of the grain is blunt, and the grain tapers towards the germ end, which is acutely pointed.

Maize

The grain of maize (see Figs. 3 and 6) is much larger than those of other cereals. The basal part, which is attached by a short stalk to the rachis, is narrow, the apex broad. The embryo, relatively large scutellum (10–13% of the grain) and the endosperm are within the pericarp and testa, which are fused to form a "hull" (corresponding morphologically with the *bran* of other cereals: the *hull* of oats, barley and rice does not correspond with the so-called "hull" of maize, being formed from the adherent lemma and palea. These parts of the maize floret are lost in threshing). The part of the hull overlying the germ is called the "tip cap". There is no ventral furrow or crease in maize.

Rice

The grain (see Figs. 3 and 6), a covered caryopsis, slightly smaller than wheat, is flattened laterally, and has a small point at the end distal from the germ. There is no ventral furrow.

The proportion of husk in the rice grain averages about 21%. Varieties of rice are classified according to kernel weight, length, and shape—which is described as round, medium or long, and defined by the ratio of the length to the breadth. Rice terms published by a consultative committee of the F.A.O., with their definitions, are shown in Table 7.

Classification according to cooking properties on the basis of final gelatinization temperature has also been suggested (Juliano *et al.*, 1964).

TABLE 7

F.A.O. RICE TERMS*

Kernel size	Length (mm)	Kernel shape	Ratio: L/B	Kernel size (wt.)	1000 kernel wt. of milled rice (g)
Extra long	>7	Slender	>3	Very large	>28
Long	6–7	Medium	2·4–3·0	Large	22–28
Middling	5–5·99	Bold	2·0–2·39	Small	<22
Short	<5	Round	<2		

* Quoted by Grist (1959).

CELLULAR STRUCTURE OF PARTS OF THE GRAIN

Pericarp

The epidermis of the caryopsis of cereals consists of thin-walled, long, rectangular cells; the hypoderm (the next layer within) is of varying thickness. The cells of the outer part of the pericarp are elongated in the length-wise direction of the grain. The cross cell layer of the inner pericarp consists of cells elongated in the transverse direction of the grain. The innermost layer of the pericarp becomes considerably torn during ripening, and is represented in the mature grain by a layer of branching hypha-like cells known as "tube cells". In wheat, the whole of the pericarp is thin and papery in the ripe grain, the outer layers often splitting off during cleaning, conditioning or milling, when they are known as "beeswing" (cf. Fig. 4).

Testa and hyaline layer

The testa or seed coat is a thin single or double layer, with the cellular structure almost obliterated. The inner layer of the testa is often deeply pigmented, and gives the grain its characteristic colour. The hyaline layer (the remains of the nucellar epidermis) is colourless and devoid of any obvious cellular structure. The hyaline layer of wheat was formerly regarded as a relatively impervious layer that impedes passage of water through the bran; it has been shown by Hinton (1955), however, that the water-impervious layer is the testa.

Aleurone

The aleurone layer in the wheat grain consists of a single layer of thick-walled cubical cells, the contents of which are devoid of starch but rich in protein and fat. In barley, the aleurone is generally two to four cell layers thick, but otherwise similar to that of wheat. In rice, it is two or three cell layers thick in the species *indica*, but five or six layers thick on the dorsal side of *japonica*. The cells are rectangular parenchyma with thin cell walls. The aleurone is a single layer of cells in rye, oats and maize.

All the layers of the pericarp, the testa and the hyaline layer closely surround the endosperm, fold in at the crease and form a loose covering over the germ. The aleurone layer, however, which is morphologically a part of the endosperm, generally surrounds the starchy endosperm only, and ceases where the latter abuts on the scutellum. In rice, however, the ventral side of the embryo is protected by a prolongation of the aleurone layer, which extends from the upper part of the embryo to its base. The aleurone is further covered by the pericarp. The aleurone layer and all the covering layers external to it comprise the "bran" (cf. Fig. 4), an important by-product in flour-milling.

Endosperm

The starchy endosperm of wheat (generally called just the "endosperm") consists of thin-walled cells which are variable in size, shape, and composition of the contents in different parts of the grain. Those adjacent to the aleurone cells (the peripheral endosperm) are small and cubical, those further in are elongated in a radial direction (prismatic endosperm cells), becoming large and polygonal (central endosperm cells) in the centre of the cheeks. The contents of the endosperm cells of each region consist mainly of starch and protein, the starch in the form of lenticular or spherical granules tightly packed together, the protein filling the intergranular spaces (cf. p. 142; see Fig. 24).

Most of the starch granules in the prismatic and central endosperm cells of wheat fall within two size ranges: large, 15–40 μ in diameter, and small, 1–10 μ in diameter, whereas those in the peripheral endosperm cells are mainly intermediate in size, viz. 6–15 μ in diameter. There is relatively more protein, and the starch granules are less tightly packed, in the peripheral endosperm cells than in the remainder of the endosperm. These differences in structure and composition between peripheral and inner endosperm cells affect the distribution of protein within the endosperm (cf. Table 16).

The texture of the endosperm of *maize* is variable according to the type of maize and the region of the kernel. The crown region of the endosperm (at the opposite end from the germ), which is light in colour, contains loosely packed starch granules with little protein, whereas the horny region (towards the base), which is more intensely coloured in yellow varieties, has smaller starch granules which are embedded in sheets of proteinaceous material. The oil and protein contents of the horny endosperm are more than double those in the crown region. In the more floury types of maize the crown region predominates: in some of these the crown region contracts during maturation, producing a noticeable indentation. Such types, varieties of *Zea mays indentata* are called *dent* maize (corn). Varieties of *Z. mays indurata*, in which the horny region predominates, are called *flint* maize (corn).

The kernels of *sorghum* resemble those of maize in having horny and starchy parts, but there is a relatively larger proportion of horny endosperm in sorghum than in maize. The outer part of the endosperm in sorghum consists of a layer of dense proteinaceous cells. The grains of sorghum are, however, much smaller than those of maize (see Table 6) and are rounded in shape.

Starch granules

The size and shape of the starch granules in the endosperm cells vary from one cereal to another; the granules in wheat, rye, barley and maize are simple, whereas those in rice are compound (see Table 8). Oats contains both compound and simple grains, with the former preponderating. Knowledge of the usual size of starch granules is of importance when applying the principles of air classification to finely ground flour (cf. p. 145).

TABLE 8

CHARACTERISTICS OF STARCH GRANULES OF CEREALS*

Cereal	Size	Shape	Notes
Wheat	Large: 15–40 μ Small: 1–10 μ	Spherical or lenticular Spherical	Granules simple
Rye	Large: 25–60 μ Small 2–10 μ	Spherical or lenticular Spherical	Concentric markings sometimes noticeable. Hilum visible
Barley	Large: 20–35 μ Small: 1–5 μ	Lenticular, kidney-shaped or sub-angular Spherical or spindle-shaped	No concentric markings Often in groups
Oats	Compound granules up to 60 μ Simple: 2–10 μ	Lenticular Spherical	Containing up to 80 individual granules Single granules
Maize	2–30 μ 2–30 μ	1. Angular, polygonal 2. Spherical	In flinty endosperm In mealy endosperm. No concentric rings. Star-shaped hilum present
Rice	Compound granules: individual granules 2–12 μ	Angular	Containing up to 150 individual granules

* Sources of data: Duly (1928), Jones *et al.* (1959), Kerr (1950), Matz (1959).

TABLE 9

PROPORTION OF PARTS IN CEREAL GRAINS*

Cereal	Hull	Pericarp + testa	Aleurone	Endosperm	Germ — Embryo	Germ — Scutellum
Wheat:						
Thatcher		8·2	6·7	81·5	1·6	2·0
Vilmorin 27		8·0	7·0	82·5	1·0	1·5
Argentinian		9·5	6·4	81·4	1·3	1·4
Egyptian		7·4	6·7	84·1	1·3	1·5
Maize:						
Flint		6·5	2·2	79·6	1·1	10·6
Sweet		5·1	3·3	76·4	2·0	13·2
Sorghum		8·0		82·0	10·0	
Rice:						
Whole grain	20	4·8		73·0	2·2	
Kernel:						
Indian		7·0		90·7	0·9 |	1·4
Egyptian		5·0		91·7	3·3	
Oats:						
Whole grain	25	9·0		63·0	1·2	1·6
Kernel		12·0		84·0	1·6	2·1
Barley:						
Whole grain	13	2·9	4·8	76·2	1·7	1·3
Kernel		3·3	5·5	87·6	1·9	1·5
Rye		10·0		86·5	1·8	1·7

* Data for wheat, maize, and rice kernel and for embryo and scutellum of oats, barley and rye from Heathcote *et al.* (1952), Hinton (1944, 1955), and Hinton and Shaw (1953). Other data for rye from Matz (1959). Other data for oats from Martin (1952). Data for sorghum from Hubbard *et al.* (1950).

Embryo

The morphological parts of the embryo or young plant are listed above, p. 23. The epiblast is well developed in rice but vestigial in other cereals. Associated with the embryo is the scutellum, a shield-shaped organ situated between the embryo and the endosperm, and considered by some to be a modified cotyledon, which has the function of mobilizing the stored food reserves in the endosperm and transmitting them to the embryo when the grain germinates. The germ and scutellum are rich in protein and fat, and the scutellum is the seat of most of the vitamin B_1 in the grain (cf. Ch. 3: Tables 16, 24, 25, 26, and Fig. 9).

Proportion of parts in the grain

The proportions, as percentages by weight, in which the various parts occur in cereal grains are shown in Table 9. For comparative purposes, the values for oats, barley and rice are shown both as proportion of the whole grain (with hull, as harvested) and of the kernel (after removal of hull).

REFERENCES

DULY, S. J. (1928), *Grain*, O.U.P., London.

GRIST, D. H. (1959), *Rice*, 3rd edition, Longmans, London.

HEATHCOTE, J. G., HINTON, J. J. C. and SHAW, B. (1952), The distribution of nicotinic acid in wheat and maize, *Proc. Roy. Soc.* **139**: 276–87.

HINTON, J. J. C. (1944), The chemistry of wheat with particular reference to the scutellum, *Biochem. J.* **38**: 214–17.

HINTON, J. J. C. (1955), Resistance of the testa to entry of water into the wheat kernel, *Cereal Chem.* **32**: 296–306.

HINTON, J. J. C. and SHAW, B. (1953), The distribution of nicotinic acid in the rice grain, *Brit. J. Nutrit.* **8**: 65–71.

HUBBARD, J. E., HALL, H. H. and EARLE, F. R. (1950), Composition of the component parts of the sorghum kernel, *Cereal Chem.* **27**: 415–20.

HUTCHINSON, J. B., KENT, N. L. and MARTIN, H. F. (1952), The kernel content of oats, *J. Nat. Inst. Agric. Bot.* **6**: 149–60.

JONES, C. R., HALTON, P. and STEVENS, D. J. (1959), Separation of flour into fractions of different protein contents by means of air classification, *J. Biochem. Microbiol. Technol. Engng.* **1**: 77–98.

JULIANO, B. O., BAUTISTA, G. M., LUGAY, J. C. and REYES, A. C. (1964), Studies on the physico-chemical properties of rice, *Agric. Fd. Chem.* **12**: 131.

KERR, R. W. (1950), *Chemistry and Industry of Starch*, New York.

MARTIN, H. F. (1952), Private communication.

MATZ, S. A. (1959) (Ed.), *The Chemistry and Technology of Cereals as Food and Feed*, Avi Publ. Co., Westport, Conn., U.S.A.

FURTHER READING

BARLING, D. M., *An Introduction to Cereal Structure and Varietal Identification*, Inst. Corn Agric. Merch. Ltd., London, 1963.

BEAVEN, E. S., *Barley*, Duckworth, London, 1947.

CARLETON, M. A., *The Small Grains*, Macmillan, New York, 1920.

HUNTER, H., *The Barley Crop*, Crosby Lockwood, London, 1928.

HUNTER, H., *Crop Varieties*, Spon, London, 1951.

KENT, N. L. and JONES, C. R., The cellular structure of wheat flour, *Cereal Chem.* **29**: 383, 1952.

PERCIVAL, J., *The Wheat Plant*, Duckworth, London, 1921.

CHAPTER 3

CHEMICAL COMPOSITION OF CEREALS

THE ripe grain of the common cereals consists of carbohydrates, nitrogenous compounds (mainly proteins), fat, mineral salts and water, together with small quantities of vitamins, enzymes and other substances, some of which are important nutrients in the human dietary.

Carbohydrates are quantitatively the most important constituents, forming about 83% of the total dry matter of wheat, barley, rye, maize, sorghum and rice, and about 79% of oats. The carbohydrates present in cereal grains include starch, which preponderates, cellulose, hemicelluloses, pentosans, dextrins and sugars. However, in proximate analysis it is customary to state the carbohydrate in two parts: the "crude fibre", which is estimated as that portion of the carbohydrate which is insoluble in dilute acids and alkalis under prescribed conditions, and the "soluble carbohydrates", which are calculated as the remainder left after accounting for crude fibre, nitrogenous compounds, fat and mineral matter. Neither crude fibre nor soluble carbohydrates are pure chemical substances; nevertheless, a knowledge of their content in the grain is of importance in relation to nutritional and digestibility studies, and is probably of greater usefulness than analytical data for the pure carbohydrate constituents.

Representative values for the proximate composition of cereal grains as harvested, and of the kernels of oats and rice, are shown in Table 10. For any one cereal, a wide range of values for each chemical constituent is encountered when a series of samples is analysed: the more diverse the types represented in the series the wider the spread of the ranges. Single figures, such as those given in Table 10, are therefore of limited value, and the intention in presenting them is only to reveal the major differences between the cereals.

TABLE 10
PROXIMATE COMPOSITION OF CEREAL GRAINS

Cereal	Moisture content (%)	Protein (N×6·25)* (%)	Fat (%)	"Soluble carbo-hydrate" (%)	Crude fibre (%)	Ash (%)	Source of data†
Wheat:							
Manitoba	(15)	13·6	2·5	63·0	2·2	1·5	1
English	(15)	8·9	2·2	66·8	2·1	1·5	1
Mixed grist	12·2	13·2	1·8	69·0	2·1	1·7	2
Maize							
Flint	11·5	9·8	4·3	71·0	1·9	1·5	3
Dent	11	9·4	4·1	72·1	2·0	1·4	4
Sweet	10·1	10·9	8·2	67·0	2·0	1·8	4
Sorghum	(11)	11·0	3·2	70·9	2·4	1·5	9
Rye	10	12·4	1·3	71·7	2·3	2·0	5
Barley	15	10	1·5	66·4	4·5	2·6	6
Rice:							
Paddy	(12)	8·0	1·9	62·7	9·0	6·3	7
Brown	(12)	9·7	2·4	73·2	1·1	1·6	7
Polished	(12)	8·6	0·4	78·2	0·3	0·5	7
Oats:							
Whole grain	11	10·3	4·7	62·1	9·3	2·6	8
Groats	11	13·3	6·2	66·4	1·2	1·9	8

* N × 5·7 for wheat and rye; N × 5·95 for rice.

† 1. McCance *et al.* (1945). 2. Booth *et al.* (1946). 3. Watt and Merrill (1950). 4. Fan *et al.* (1963). 5. Schopmeyer (1962). 6. Watson (1953). 7. Juliano *et al.* (1964). 8. Original data. 9. Hubbard *et al.* (1950).

The husked caryopses of oats, barley and rice (paddy) have a crude fibre content 2–5 times that of wheat, rye, sorghum and maize, which are naked caryopses. The protein content of rice is lower than that of all other cereals. Removal of the husk of rice and oats during processing increases the protein content of the product; de-husked (brown) rice is still comparatively low in protein content, but de-husked oats (groats) equal or exceed wheat in protein content.

Oats and maize are relatively rich in fat, and de-husked oats (groats) are particularly valuable nutritionally for their fat content.

The mineral ash content is higher in barley, oats and rice (paddy) than in maize, sorghum, wheat and rye; this is a further consequence of the presence of husk, which is rich in minerals, around the grains of the former group of cereals (cf. Table 22). When comparison is made among the cereals in the same morphological condition, viz. after shelling the husked types, the differences in mineral ash content are much reduced.

EFFECT OF PROCESSING ON CHEMICAL COMPOSITION OF CEREAL PRODUCTS

The processing of cereals for the manufacture of food may bring about alteration in the chemical composition in a number of ways:

1. Parts of the grain may be separated during processing and removed from the product, or the product may constitute only a fraction of the grain.

2. The various nutrients may be distributed non-uniformly throughout the various parts of the grain, so that when separation is made (as in 1) certain nutrients are preferentially lost from or concentrated into the product.

3. The processing treatments may bring about changes in the nutrients themselves: these may be chemical changes as, for example, when enzymes are inactivated by steam treatment (cf. p. 216), or changes in distribution, e.g. when translocation of vitamins is brought about during the parboiling of rice (cf. p. 240).

Distribution of nutrients in wheat

Two methods have been extensively used to study the general

pattern of distribution of nutrients within the cereal grains:
(1) analysis of milling fractions, and (2) analysis of the morphological parts of the grain, obtained by manual or mechanical
dissection.

Milling fractions generally do not correspond precisely with
particular morphological parts of the grain; moreover, their
composition is variable from mill to mill. Analysis of milling
fractions thus gives only an indication of the path followed by
any nutrient at each stage in the processing.

A clearer picture of the distribution of nutrients can be
obtained by analysis of the dissected morphological parts of the
grain, and the effects of milling or processing treatments, in
which parts of the grain are separated or discarded, can be
foreseen.

Starch is present only in the endosperm, but protein occurs
throughout the grain (cf. pp. 44 and 162). The fibre is restricted
almost entirely to the bran, with only about 10% of the total
fibre in the endosperm and germ. About half of the total fat is in
the endosperm, about one-fifth in the germ, the remainder in
the bran, but more in the aleurone than in the pericarp and
testa. Distribution of ash resembles that of fibre, with over half
the total in pericarp, testa and aleurone. Table 11 summarizes
the facts presented in this paragraph.

TABLE 11

PERCENTAGE OF THE TOTAL CONSTITUENTS OF WHEAT PRESENT IN THE
MAIN MORPHOLOGICAL PARTS*

Part	Wt. (g per 100 g of grain)	Constituents				
		Starch	Protein	Fibre	Fat	Ash
Pericarp, testa, aleurone	15	0	20	70	30	67
Endosperm	82	100	72	8	50	23
Embryo, scutellum	3	0	8	3	20	10

* Data from Hinton (1952, 1959).

Distribution of nutrients in maize

The proportions of the constituents in the main morphological parts of maize, shown in Table 12, are similar to those for wheat as regards starch, but show a larger proportion of protein, fat and ash in the germ, with less in the bran. This arises partly from the considerably larger contribution made by the germ to the weight of the grain in maize than in wheat.

TABLE 12

PERCENTAGE OF THE TOTAL CONSTITUENTS OF MAIZE PRESENT IN THE
MAIN MORPHOLOGICAL PARTS*

Part	Wt. (g per 100 g of grain)	Constituents			
		Starch	Protein	Fat	Ash
Bran	5	0	2	1	2
Endosperm	82	98	75	15	17
Germ, tip cap	13	2	23	84	81

* Data derived from Shollenberger and Jaeger (1943).

In barley, the husk, which contributes 13% on average to the weight of the grain, contains 64% of the total fibre, 32% of the ash. The husk of oats (25% by wt. of average) and of rice (20% by wt.) contains 85–90% of the total fibre, 40% and 79%, respectively, of the total ash, and only 4–9% of the total protein and fat.

CARBOHYDRATES

Starch

Starch is the most important carbohydrate in all the cereals, constituting about 60% of the entire wheat grain, and 70–71% of the endosperm of wheat. Starch is a glucose polymer based on α-linkages, mostly in the 1 : 4 position and a few in the 1 : 6 position. It occurs in two forms: straight-chained amylose, and branched-chained amylopectin. About 23% of the starch is amylose in wheat, barley and oats, about 27% in rice and dent maize, about 50% in amylo-maize, but in waxy maize all the starch is amylopectin. The unit chain of amylose contains 20–30 glucose units, that of amylopectin 18–36 units (Greenwood, 1956).

Starch is insoluble in cold water; when heated with water it absorbs water and swells: this process is known as gelatinization.

The starch occurs as granules (cf. p. 30), which are susceptible to mechanical damage during grinding and milling; the damaged starch plays an important role in the baking process (cf. p. 175).

Cellulose (fibre)

Cellulose is the main constituent of the cell wall of cereal grains, and forms the bulk of the "crude fibre". It is a glucose polymer with the same empirical formula as starch, but based on the much more stable β-linkage. The fibre content of the whole wheat grain is about 2%; in the endosperm it is about 0·1%, in bran 12–14%.

Sugars

The free sugar content of wheat is about 2·5% (Fraser, 1958). The sugars extractable by 80% alcohol from the flours of Canadian wheat and rye, as estimated by paper chromatography, are shown in Table 13.

TABLE 13
SUGARS IN WHEAT AND RYE FLOURS*

Sugar	HRS wheat (%)	Durum wheat (%)	Rye (%)
Glucose	0·01–0·09	0·02–0·04	0·05
Fructose	0·02–0·08	0·04–0·09	0·06
Sucrose	0·19–0·26	0·26–0·57	0·41
Maltose	0·06–0·10	0·10–0·15	0·14
Oligosaccharides	1·26–1·31	0·67–1·05	2·03

* Data from Vaisey and Unrau (1964).

The oligosaccharides were maltotriose, tetraose, and pentaose, yielding glucose on hydrolysis. Raffinose (0·07%) has been reported in wheat flour by Koch *et al.* (1951). Dextrins, compounds intermediate between starch and sugar, are also present in flour.

The role of sugars in breadmaking will be discussed in Chapter 10.

Sugars are of considerable importance in the malting of barley (cf. p. 200); the range of values for the contents of individual sugars in barley is shown in Table 14.

TABLE 14
SUGARS IN BARLEY *

Sugar	mg per 100 g barley (d.m.b.)
Glucose	20–93
Fructose	33–159
Maltose	0–135
Sucrose	343–1690
Raffinose	144–832
Gluco-di-fructose	70–433
Fructosans:	
Ethanol-soluble	97–536
Water-soluble	40–900

* Data from Harris (1962).

PROTEINS (cf. Ch. 9, p.158)

In their primary structure, protein molecules consist of chains of amino acids linked together by peptide bonds between the carboxyl (COOH) group of one amino acid and the alpha-amino (NH_2) group of the next.

Some eighteen different amino acids occur in the proteins of cereals. The proportions in which they occur, and their order in the chains, give the characteristics of particular proteins. The main peptide chain or backbone formation of the protein molecule may be linked to adjacent molecules by disulphide bonds from cystine residues (secondary structure). The peptide chains may be coiled in spirals, with hydrogen bonds linking the protruding side chains (tertiary or alpha-helix main-chain conformation).

Types of protein

Osborne (1907) classified the proteins of flour, according to solubility characteristics, into five categories (Table 15), but research over the last decade has shown that this is an over-simplification. Each of the groups is highly heterogeneous and contains many individual protein species (cf. Pence and Nimmo, 1964).

TABLE 15
PROTEINS IN WHEAT (FROM OSBORNE)

Type of protein	Approx. % of total protein	Approx. % of wheat	Extracted by
Albumin	2·5	0·3 ⎫	Dilute salt
Globulin	5·0	0·6–0·7 ⎬	solutions
Proteose	2·5	0·3 ⎭	
Prolamin (gliadin)	40–50	4·0	70% alcohol
Glutelin (glutenin)	40–50	4·0	Dilute acids and alkalis

Pence *et al.* (1954) give the following proportions for the main constituents of flour proteins (in five varieties of wheat, ranging 6–14% in flour protein content): albumin 6–12%, globulin 5–11%, gluten 78–85%.

The albumins and globulins in flour are commonly referred to as the soluble proteins. The albumin proteins are responsible for part of the differences in baking characteristics among flours (Pence *et al.*, 1951), and the globulins also may be essential for proper baking performance.

The insoluble portion is made up of numerous components. The portion soluble in aqueous alcohol is still called gliadin for convenience, but it is known to contain at least eight components (Woychik *et al.*, 1961). Jones *et al.* (1961) give 42,000–47,000 for the molecular weight of gliadin. The residue, insoluble in alcohol, is called glutenin. Nielsen *et al.* (1962) consider that it consists of units of molecular weight about 20,000, which are bonded together by disulphide bonds into macro-units with molecular weight going up into the millions. The gliadin and glutenin form, with water and salts, the substance gluten when a flour-water dough is kneaded (cf. pp. 147 and 173). Characterization of proteins by their solubility has been superseded by characterization dependent on electrophoretic and sedimentation properties.

The gluten complex has elasticity and flow properties of unique value for the baking of bread and other products. The elastic properties, which are developed during mixing, appear to involve sulphydryl groups, possibly their oxidation to disulphide bonds, possibly the formation of new bonds. McDermott and Pace (1959) solubilized flour protein by splitting the disulphide bonds, to form S-sulphonic acid groups; they concluded that the rigidity of the insoluble fraction is partly due to the disulphide cross-linked structure. Grosskreutz (1961) has proposed a model of gluten sheet structure to explain the flow property of gluten (see Fig. 8); slip-planes are provided by lipoprotein leaflets interspersed between protein platelets of polypeptide chains (see Pence *et al.*, 1964).

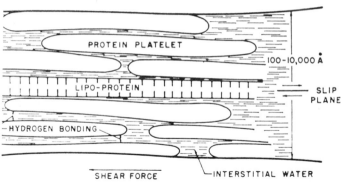

FIG. 8. Model of gluten sheet structure proposed by Grosskreutz. (Reproduced from J. C. Grosskreutz, *Cereal Chem.* **38**: 336, 1961, by courtesy of the author, and the Editor of *Cereal Chemistry*.)

Distribution of protein in the grain

Protein is found in all the tissues of cereal grains, higher concentrations occurring in the embryo, scutellum and aleurone layer than in the endosperm, pericarp and testa. Within the endosperm, the concentration of protein increases from the centre to the periphery (cf. p. 162). The protein concentration in the dissected morphological parts of wheat (variety: Vilmorin 27) and maize grains, and the proportion contributed by each part to the total protein of the grain, are shown in Table 16.

TABLE 16

DISTRIBUTION OF PROTEIN IN WHEAT AND MAIZE*

Part of grain	Proportion of kernel		Protein content † (N × 6·25)		Proportion of total nutrient in kernel	
	Wheat (%)	Maize (%)	Wheat (%)	Maize (%)	Wheat (%)	Maize (%)
Pericarp	8	6·5	4·4	3·0	4·0	2·2
Aleurone	7	2·2	19·7	19·2	15·5	4·7
Endosperm	(82·5)	(79·6)	—	—	(72·5)	(71·0)
outer	12·5	3·9	13·7	27·7	19·4	11·9
middle	12·5	58·1	8·8	7·5	12·4	48·2
inner	57·5	17·6	6·2	5·6	40·7	10·9
Embryo	1	1·1	33·3	26·5	3·5	3·2
Scutellum	1·5	10·6	26·7	16·0	4·5	18·9

* Data from Hinton (1953).

† 14·5% m.c. basis.

The aleurone layer contributes a smaller proportion to the total grain weight in maize than in wheat, and contains less of the total protein, although the concentration in that tissue is similar in both cereals. The scutellum is relatively larger in maize than in wheat and, although the protein concentration therein is lower, the maize scutellum contributes a considerably larger proportion of protein to the total than does wheat scutellum. The endosperm, about 80% of the grain by weight, contributes 70% of the protein in each cereal.

Amino acid composition

The amino acid composition of the total protein of flour milled from Manitoba wheat, English wheat (variety: Hybrid 46), and a commercial mixed grist is shown in Table 17.

TABLE 17

AMINO ACID CONTENT OF VARIOUS WHEAT FLOURS*
(G AMINO ACID N/100 G OF TOTAL N)

Amino acid	Manitoba	English	Mixed grist
Alanine	2·86	3·00	2·87
Arginine	6·99	7·44	6·72
Aspartic acid	2·65	2·67	2·66
Cystine	1·74	1·88	1·68
Glutamic acid	19·56	20·31	19·94
Glycine	3·84	3·74	3·93
Histidine	3·89	3·93	3·59
*Iso*leucine	2·52	2·58	2·55
Leucine	4·69	4·90	4·81
Lysine	2·46	2·64	2·36
Methionine	1·12	1·13	1·05
Phenylalanine	2·72	2·88	2·80
Proline	9·19	9·02	8·83
Serine	3·81	4·09	3·89
Threonine	2·06	2·15	2·05
Tryptophan	—	—	0·98
Tyrosine	1·45	1·39	1·61
Valine	3·12	3·21	3·28
Ammonia	22·25	22·07	—
Nitrogen, % d.m.b.	2·31	1·68	2·29

* Data from McDermott and Pace (1960).

The high contents of glutamic acid (probably present in the intact protein as glutamine) and of proline and the low content of lysine are notable.

The flours from Manitoba and English wheats differ greatly in protein content and in physical characteristics exhibited by the proteins, yet the amino acid compositions of both are very similar. The figures for lysine, arginine and cystine, however, do show some differences which, although small, are important. Moreover, the amino acid composition of wheat protein is not constant over a wide range of protein contents. In particular, the amino acid lysine forms a smaller proportion of the protein in wheats of higher protein contents. This was found by McDermott and Pace

(1960) when comparisons were made between (i) varieties of varying protein content, (ii) vitreous grains of high protein content and mealy grains of low protein content in a single wheat sample, and (iii) vitreous and mealy portions of the endosperm of piebald grains—see Table 18.

TABLE 18

NITROGEN AND LYSINE CONTENTS OF VARIOUS TYPES OF WHEAT ENDOSPERM*

Wheat	Type of endosperm	N content d.m.b. (%)	Lysine content (g amino acid N/100 g total N)
Manitoba	Total	1·55	2·98
Hybrid 46	Total	1·07	3·38
Arletta	Total	0·84	3·80
Hybrid 46	Vitreous	1·60	2·79
Hybrid 46	Mealy	1·01	3·75

* Data from McDermott and Pace (1960).

The biological value of the protein in germ and aleurone is higher than that of the endosperm proteins; the lysine content is about 2·5 times as large in the protein of these tissues as it is in that of endosperm (Stevens *et al.*, 1963).

In barley, similar types of protein have been reported: albumin and globulin, 2% of the grain, a prolamin (hordein) 4%, and a glutelin 4·5%. The hordein and glutelin, however, differ from wheat gliadin and glutenin in not forming gluten when washed out. Immature grains of wheat and barley contain relatively large proportions of salt-soluble and glutelin fractions, and relatively less prolamin, but as maturation proceeds the proportion of the prolamin increases more steeply than that of the other fractions. The relative proportions are also related to the total nitrogen content of the wheat and barley (Bishop, 1928; 1937). Samples (of a particular variety) of barley with high total nitrogen content contained relatively more prolamin and less salt-soluble nitrogen than those of lower total protein content.

Technology of Cereals

Rye contains a prolamin which appears to be identical with the gliadin of wheat.

Table 19 gives data for the amino acid contents of the cereals. Rice and oats surpass the other cereals in their content of arginine; maize and sorghum are characterized by low tryptophan and high leucine contents.

TABLE 19

AMINO ACID CONTENT OF CEREAL GRAINS*

(G AMINO ACID/16 G NITROGEN)

Amino acid	Wheat†	Barley§	Rye§	Oats§	Rice‡	Maize§	Sorghum
Arginine	4·3	5·0	5·0	6·6	7·7	5·0	4·7
Cystine	2·1	2·1	1·8	1·8	1·1	2·1	—
Histidine	2·1	1·9	2·1	1·9	2·3	2·4	3·3
*Iso*leucine	3·8	3·8	3·9	4·6	3·9	4·0	4·7
Leucine	6·4	6·9	6·1	7·0	8·0	12·0	14·3
Lysine	2·7	3·4	3·7	3·7	3·7	3·0	2·9
Methionine	1·6	1·4	1·6	1·4	2·4	2·1	1·6
Phenylalanine	4·6	5·0	4·6	5·0	5·2	5·0	4·3
Threonine	2·9	3·7	3·6	3·4	4·1	4·2	3·8
Tryptophan	1·3	1·4	1·3	1·3	1·4	0·8	0·7
Tyrosine	3·2	3·5	4·2	3·8	3·3	3·8	2·7
Valine	4·3	5·0	5·0	5·4	5·7	5·6	6·0
Alanine	3·4	4·5	—	5·1	6·0	9·9	—
Aspartic acid	5·0	5·9	—	4·2	10·4	12·3	—
Glutamic acid	27·7	20·5	19·7	18·4	20·4	15·4	21·9
Glycine	3·8	4·3	—	4·2	5·0	3·0	—
Proline	10·1	9·3	—	5·8	4·8	8·3	—
Serine	4·8	3·7	3·8	3·4	5·2	4·2	—

* Data for rice (except tryptophan) from Juliano *et al.* (1964); data for sorghum from Bressani and Rios (1962); all other data calculated from Hughes (1960).

† The original data are given for flour (100% extraction).

§ Whole grains.

‡ Brown rice—whole grains, no hulls; protein content 11·1% (d.m.b.).

FAT

The fat content of wheat, rye, barley and rice is 1–2%, that of sorghum 3%, that of maize and whole oats 4–6%. As the husk of oats contains a negligible amount of fat, the fat content of oat kernels is higher still, within the range 5–10%, average 7%.

In wheat, the germ contains 6–11% of fat, the bran 3–5%, and the endosperm 0·8–1·5%. In maize, the germ is even richer in fat, containing 35%, but the bran contains somewhat less, only about 1%.

The fat of cereals consists of the glycerides of fatty acids; some figures available from the literature for the composition of the fatty acids obtained from the fats of cereals are shown in Table 20.

Cereals also contain phospholipids. Lecithin, consisting of one molecule of glycerol linked to two molecules of fatty acid and one molecule of phosphoric acid, which is, in turn, linked to choline, is an example of this class of compounds. The oil of cereals contains up to 4% of phospholipids. Sugar-containing lipids have been found in the oil of wheat endosperm.

The fat in milled cereal products is liable to undergo two types of deterioration: hydrolysis, by the action of the enzyme lipase which is present in the grain, and oxidation, which may occur enzymically, by action of the enzyme lipoxidase, or non-enzymically, in the presence of oxygen. Normally the enzymes and the fat do not come into contact in the intact grain, but damage to the germ and the fragmentation that occurs in milling may bring the fat and the enzymes together, thereby promoting deterioration. The products of fat hydrolysis are glycerol and free fatty acids; sound whole grains normally contain small amounts of the free acids (e.g. 4–10% of the fat in oat kernels), but larger amounts, due to damage and deterioration, give rise to unpleasant flavours. The products of fat oxidation cause the odour and flavour of rancidity. For further discussion of fat deterioration, see pp. 190 and 213.

Natural antioxidants have recently been found in oats (cf. p. 223).

The germ is separated from the endosperm in white flour milling in order to improve the keeping quality of the flour.

TABLE 20

COMPOSITION OF THE FATTY ACIDS OF CEREAL LIPIDS*

Fatty acids	Wheat			Barley (%)	Rye (%)	Oats (%)	Rice (%)	Maize germ (%)
	Grain (%)	Germ (%)	Endo-sperm (%)					
Saturated								
$C_{14:0}$ Myristic	0·1	—	—	1·0	—	—	—	—
$C_{16:0}$ Palmitic	24·5	18·5	18·0	11·5	21·0	10·4	17·6	7·8–10·2
$C_{18:0}$ Stearic	1·0	0·4	1·2	3·1		—		0·9–3·5
Unsaturated								
$C_{16:1}$ Palmitoleic	0·8	0·7	1·0	—	—	—	—	—
$C_{18:1}$ Oleic	11·5	17·3	19·4	28·0	18·0	58·5	47·6	23·5–49·6
$C_{18:2}$ Linoleic	56·3	57·0	56·2	52·3	61·0	31·1	34·0	34·3–60·8
$C_{18:3}$ Linolenic	3·7	5·2	3·1	4·1	—	—	0·8	—
Others and unsapon.	1·9	0·8	1·1	—	—	—	—	0·3–3·0

* Sources of data:—Wheat: Nelson *et al.* (1963). Barley: McLeod and White (1961). Rye: Matz (1959). Oats: Amberger and Wheeler-Hill (1927). Rice: Mickus (1959). Maize germ: various sources quoted by Hilditch (1956).

Separation of the maize germ, which has a much higher fat content than that of wheat germ, is equally important in the manufacture of maize grits and corn flour (cf. p. 245).

MINERAL MATTER

About 95% of the mineral matter of the cereals which are naked caryopses (viz. wheat, sorghum, rye and maize) and of the kernels of oats, barley and rice consists of the phosphates and sulphates of potassium, magnesium and calcium. The potassium phosphate is probably present in wheat mainly in the form of KH_2PO_4 and K_2HPO_4. Some of the phosphorus is present as phytic acid (cf. p. 163). The reported values for the contents of sodium, chlorine and sulphur are somewhat variable. Important minor elements are iron, manganese and zinc, present at a level of 1–5 mg per 100 g, and copper, about 0·5 mg per 100 g. Besides these, a large number of other elements are present in trace quantities. Representative data from the literature are collected in Table 22.

The husk of barley, oats and rice has higher ash content than that of the kernels, and the ash is particularly rich in silica: see Table 21.

TABLE 21
ASH AND SILICA IN CEREAL HUSKS

Material	Ash content (%)	Silica in ash (%)	Source of data*
Barley husk	6·0	65·8	1
Oat husk	5·2	68·0	1
Rice husk	22·6	95·8	2

* 1. Original data. 2. Nelson *et al.* (1950).

VITAMINS

The average contents of the vitamin B constituents of cereals (whole grains) as reported in the literature are shown in Table 23.

TABLE 22

REPRESENTATIVE FIGURES FOR MINERAL CONSTITUENTS OF CEREAL GRAINS* (MG PER 100 G OF DRY MATTER)

Element	Wheat (mixed)	Barley	Rye	Oats		Maize	Rice		Sorghum
				Whole	Kernels		Paddy	Brown	
Main: K	453	580	412	460	380	339	342	118	330
P	380	440	359	341	400	322	285	290	445
S	196	160	146	199	185	151	—	—	171
Mg	157	180	92	143	120	121	90	47	105
Cl	76	120	60	100	70	45	23	286	54
Ca	51	50	31	95	66	29	68	67	22
Na	24	77	26	87	50	36	78	54	7·5
Si	12	420	6	690	28	—	1790	—	4
Minor: Fe	5	5	2·7	7	4·2	3·6	—	3·2	8
Zn	5	—	—	2·7	—	—	—	—	9
Mn	4	2	2·5	5	4	0·7	5·6	1·7	5
Cu	0·7	0·5	0·6	0·4	0·5	0·5	0·3	0·4	0·08
Trace: Ba	0·8	—	—	—	—	—	1·2	—	—
Br	0·6	—	—	—	—	—	—	—	—

B	0·5	—	—	0·12	0·12	—	0·9	—	1
Li	0·5	—	—	—	—	—	0·3	—	—
Al	0·3	—	—	0·5	—	—	—	—	0·4
Sr	—	—	*	—	—	—	0·17	—	—
Ni	0·14	0·02	—	0·2	—	—	0·11	—	0·08
Sn	0·11	—	—	—	—	—	0·015	—	0·04
Ti	0·085	—	—	—	0·04	—	—	—	0·04
F	0·07	—	—	0·04	0·04	—	0·6	—	—
Pb	0·04	—	—	—	—	—	>0·003	—	1
Mo	0·026	0·04	—	0·04	—	—	0·027	—	0·1
Co	0·003	<0·005	—	0·002	0·002	—	<0·003	—	—
I	0·014	0·002	—	0·011	0·006	0·036	—	0·002	—
As	0·01	—	—	0·05	—	—	—	—	—
Ash content, %	1·5	2·0	1·7	3·4	2·0	3·4	5·9	0·9	1·5

N.B. A blank in the table indicates that no reliable information has been found.

* Sources of data: Bergner and Wagner (1960, 1961). Bridges (1935). Connolly and Maguire (1963). Cooke (1962). Fournier and Digaud (1948). Kent (1946). Kent-Jones and Amos (1957). McCance et al. (1945). McNeill (1952). Meyer et al. (1950). Moir (1946). Muir (1959). Okamura et al. (1961). Pedrero (1962). Schmorl (1933). Sherman (1941).

TABLE 23

VITAMIN CONTENTS OF THE CEREAL GRAINS*

(μG/G)

Vitamin	Wheat		Barley	Rye	Oats (whole)	Rice (brown)	Maize	Sorghum
	Manitoba	English						
Vitamin B$_1$	4	2·9	6·5	4·6	5·7	4·0	4·5	3·1
Riboflavin (B$_2$)	1·2	1·1	1·2	1·5	1·3	0·6	0·9	1·1
Nicotinic acid	70	50	115	10	9·4	53	23	51
Pantothenic acid	10–15		4·4	10·4	9	17	4·6	7
Biotin	0·1			0·06			0·1	0·3
Pyridoxin (B$_6$)	5		11·5	3·3	1·2	10·3	6·9	6·4
Folic acid	0·5			0·3	0·2			
Choline	1000		1100		980			
Inositol	2500							
p-Aminobenzoic acid	1							
Vitamin B$_{12}$	0·001–0·002							

* Sources of data include: Andrews *et al.* (1942). Anon. (1958). Aykroyd and Swaminathan (1940). Barton-Wright (1945). Fournier and Digaud (1948). Frey and Watson (1950). Grist (1959). Hall *et al.* (1952). Hinton (1948). Hodson and Norris (1939). Horder *et al.* (1954). Ihde and Schuette (1941). Kent-Jones and Amos (1957). Kik and Landingham (1943). McCance *et al.* (1945). McCance and Widdowson (1960). Moir (1946). Nelson (1953). Schultz *et al.* (1941). Sebrell Jr. and Harris (1954). Shaw (1953). Thomas *et al.* (1942). Williams *et al.* (1943).

Variation from one cereal to another is remarkably small except for nicotinic acid (niacin), the concentration of which in barley, wheat, sorghum and rice is relatively much higher than in maize, rye and oats.

The vitamin E content of wheat is given as about 25 μg/g (Horder *et al.*, 1954). (See also tocopherols on p. 58.)

Distribution of vitamins in the wheat grain

The principal vitamins of the B group—aneurin or thiamine (vitamin B_1), nicotinic acid (niacin), riboflavin (vitamin B_2), pantothenic acid, and pyridoxin (vitamin B_6)—are distributed non-uniformly throughout the grain. Details of the distribution have been worked out by Hinton and his associates, who dissected wheat grains into their morphological parts and assayed them for their vitamin contents. Their results for wheat are shown in Table 24.

TABLE 24

DISTRIBUTION OF B-VITAMINS IN THE WHEAT GRAIN*

Part of grain	(Variety: Vilmorin 27) Vitamin B_1	(Variety: Thatcher)			
		Nicotinic acid	Riboflavin	Pyridoxin	Pantothenic acid
(a) Concentration in the parts of the grain (μg/g)					
Pericarp, testa and hyaline	0·6	25·7	1·0	6·0	7·8
Aleurone layer	16·5	741	10	36	45·1
Endosperm	0·13	8·5	0·7	0·3	3·9
Embryo	8·4	38·5	13·8	21·1	17·1
Scutellum	156	38·2	12·7	23·2	14·1
Whole grain	3·75	59·3	1·8	4·3	7·8
(b) Distribution among the parts as percentages of total in grain					
Pericarp, testa and hyaline	1·0	4	5	12	9
Aleurone layer	32	82	37	61	41
Endosperm	3	12	32	6	43
Embryo	2	1	12	9	3
Scutellum	62	1	14	12	4

* Sources of data: Clegg (1958). Clegg and Hinton (1958). Heathcote *et al.* (1952). Hinton (1947). Hinton *et al.* (1953).

The distribution of these vitamins and of protein in the wheat grain is also shown diagrammatically in Fig. 9.

Vitamin B_1 is concentrated in the scutellum, nicotinic acid in the aleurone layer. Riboflavin and pantothenic acid are more uniformly distributed. Pyridoxin is concentrated in aleurone and germ, with very little in the endosperm. The proportion of the total vitamin contained in the various tissues of the grain is shown in Table 25 for nicotinic acid and vitamin B_1 in wheat, rice and maize.

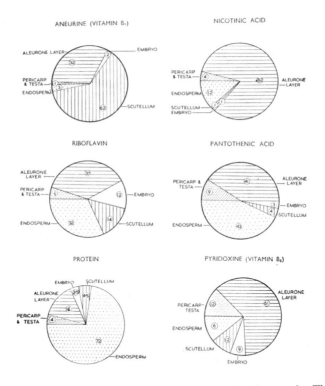

FIG. 9. Distribution of vitamins and protein in the wheat grain. The figures show the percentage of the total nutrients in the grain found in various anatomical parts. (Based on microdissections by J. J. C. Hinton. From *The Research Association of British Flour-Millers, 1923–60.*)

TABLE 25

DISTRIBUTION OF NICOTINIC ACID AND VITAMIN B$_1$ IN WHEAT,
RICE AND MAIZE*

Part of grain	Percentage of total nicotinic acid				Percentage of total B$_1$	
	Wheat (English†)	Rice (Indian)	Maize		Wheat (English†)	Rice
			Flint	Sweet		
Pericarp, testa and hyaline	4	5	2	3	} 33	34
Aleurone layer	82	80·5	63	59		
Endosperm	12	12·3	20	26	3	8
Embryo	1	0·6	2	2	2	11
Scutellum	1	1·6	13	10	62	47

* Sources of data: Heathcote *et al.* (1952); Hinton and Shaw (1953).
† Variety Vilmorin 27.

The distribution of nicotinic acid in rice and maize resembles that in wheat; the distributions of vitamin B$_1$ in rice and wheat are also quite similar. Distributions for barley, rye and oats are not known in such detail, but for vitamin B$_1$ the concentration and the proportion of the total in the germ and scutellum have been determined—see Table 26.

The proportion of the total vitamin B$_1$ in the scutellum is very high in rye and maize; it is somewhat less in barley, rice and wheat, and relatively low in oat kernels. The embryo of rice,

TABLE 26

VITAMIN B$_1$ IN EMBRYO AND SCUTELLUM OF CEREAL GRAINS*

Cereal	Wt. of tissue (g/100 g grain)		Vit. B$_1$ concn. (μg/g)		Propn. of total vit. B$_1$ of grain	
	Embryo	Scutellum	Embryo	Scutellum	Embryo	Scutellum
Wheat (average)	1·2	1·54	12·0	177	3	59
Barley (husk removed)	1·85	1·53	15	105	8	49
Rye	1·8	1·73	6·9	114	5	82
Oats (kernels)	1·6	2·13	14·4	66	4·5	28
Rice (brown)	1·0	1·25	69	189	11	47
Maize	1·15	7·25	26·1	42	8	85

* Sources of data: Hinton (1944, 1948).

which has a relatively high concentration of vitamin B_1, contains over one-tenth of the total in the grain, a larger proportion than that found in the other cereals.

The uneven distribution of the B vitamins throughout the grain is responsible for considerable differences in vitamin content between the whole grains and the milled or processed products; for further discussion, see Chapters 9, 12, 13 and 14.

TOCOPHEROLS

Wheat contains α, β, ε and ζ tocopherols, the total tocopherol content being about 3·4 mg/100 g. The biological vitamin E activities of β, ε and ζ tocopherols are 30%, 7·5% and 40%, respectively, of that of the α-tocopherol. The total tocopherol contents of germ, bran and 80% extraction flour are about 30, 6 and 1·6 mg/100 g, respectively (sources of data quoted by Moran, 1959); α-tocopherol predominates in germ, ε-tocopherol in bran and endosperm, giving α-equivalents of 65%, 20% and 35% for the total tocopherols of germ, bran and 80% flour, respectively.

The total tocopherol content of oats is 1·7–3·2 mg/100 g, d.m.b. (Brown, 1953); that of oat oil 0·6 mg/g (Green *et al.*, 1955).

REFERENCES

AMBERGER, K. and WHEELER-HILL, E. (1927), The composition of oat oil, *Z. Untersuch. Lebensm.* **54**: 417.

ANDREWS, J. S., BOYD, H. M. and TERRY, D. E. (1942). The riboflavin content of cereal grains and bread and its distribution in products of wheat milling, *Cereal Chem.* **19**: 55.

ANON. (1958), Northern Regional Research Laboratory, U.S. Dept. Agric. Quoted by MATZ, S. A. (1959).

AYKROYD, W. R. and SWAMINATHAN, M. (1940), The nicotinic acid content of cereals and pellagra, *Indian J. Med. Res.* **27**: 667.

BARTON-WRIGHT, E. C. (1945), The theory and practice of the microbiological assay of the vitamin B complex, together with the assay of selected amino acids and potassium, *Analyst* **70**: 283.

BERGNER, K. G. and WAGNER, K. (1960 and 1961), Mineral substances in oats and their distribution in the production of oat flakes, *Getreide u. Mehl* **10**: 81; **11**: 13, 61.

BISHOP, L. R. (1928), The Barley proteins. I. The composition and quantitative estimation of barley proteins, *J. Inst. Brewing* **34:** 101.

BISHOP, L. R. (1937), The importance of proteins in brewing, *J. Soc. Chem. Ind.* **56:** 244T.

BOOTH, R. G., CARTER, R. H., JONES, C. R. and MORAN, T. (1946), Cereals as food. Chemistry of wheat and wheat products, in: BACHARACH, A. L. and RENDLE, T. (1946) (Eds.), *The Nation's Food*, Soc. Chem. Ind., London.

BRESSANI, R. and RIOS, B. J. (1962), The chemical and essential amino-acid composition of twenty-five selections of grain sorghum, *Cereal Chem.* **39:** 50.

BRIDGES, M. A. (1935), *Food and Beverage Analyses*, Kimpton, London.

BROWN, F. (1953), Tocopherol content of farm feeds, *J. Sci. Fd. Agric.* **4:** 161.

CLEGG, K. M. (1958), The microbiological determination of pantothenic acid in wheaten flour, *J. Sci. Fd. Agric.* **9:** 366.

CLEGG, K. M. and HINTON, J. J. C. (1958), The microbiological determination of vitamin B_6 in wheat flour, and in fractions of the wheat grain, *J. Sci. Fd. Agric.* **9:** 717.

CONNOLLY, J. F. and MAGUIRE, M. F. (1963), An improved chromatographic method for determining trace elements in foodstuffs, *Analyst* **88:** 125.

FAN, L., CHU, P. and SHELLENBERGER, J. A. (1963), Diffusion of water in kernels of corn and sorghum, *Cereal Chem.* **40:** 303.

FOURNIER, P. and DIGAUD, A. (1948), Chemical composition of sorghum grain (*Sorghum vulgare* Pers.), *Bull. Soc. Sci. Hyg. Aliment*, **36:** 33.

FRASER, J. R. (1958), Flour survey 1950–6. *J. Sci. Fd. Agric.* **9:** 125.

FREY, K. J. and WATSON, G. I. (1950), Chemical studies on oats. I. Aneurin, niacin, riboflavin and pantothenic acids, *Agron. J.* **42:** 434.

GREEN, J., MARCINKIEWICZ, S. and WATT, P. R. (1955), The determination of tocopherols by paper chromatography, *J. Sci. Fd. Agric.* **6:** 274.

GREENWOOD, C. T. (1956), Physical chemistry of starch, *Advanc. Carbohyd. Chem.* **11:** 335.

GRIST, D. H. (1959), *Rice*, 3rd edition, Longmans, London.

GROSSKREUTZ, J. C. (1961), A lipoprotein model of wheat gluten structure, *Cereal Chem.* **38:** 336.

HALL, H. H., CURTIS, J. J. and SHEKLETON, M. C. (1952), The B-complex vitamin composition of corncobs, *Cereal Chem.* **29:** 156.

HARRIS, G. (1962), The structure and chemistry of barley and malt, in COOKE, A. H. (1962) (Ed.), *Barley and Malt—Biology, Biochemistry, Technology*, Academic Press, New York and London.

HEATHCOTE, J. G., HINTON, J. J. C. and SHAW, B. (1952), The distribution of nicotinic acid in wheat and maize, *Proc. Roy. Soc.* B **139:** 276.

HILDITCH, T. P. (1956), *The Chemical Composition of Natural Fats*, 3rd edition Chapman & Hall, London.

HINTON, J. J. C. (1944), The chemistry of wheat with particular reference to the scutellum, *Biochem. J.* **38:** 214.

HINTON, J. J. C. (1947), The distribution of vitamin B_1 and nitrogen in the wheat grain, *Proc. Roy. Soc.* B **134:** 418.

HINTON, J. J. C. (1948), The distribution of vitamin B_1 in the rice grain, *Brit. J. Nutrit.* **2:** 237.

HINTON, J. J. C. (1952), The structure of cereal grains, in BATE-SMITH, E. C. and MORRIS, T. N. (1952) (Ed.), *Food Science: A Symposium on Quality and Preservation of Foods*, Cambridge University Press.

HINTON, J. J. C. (1953), The distribution of protein in the maize kernel in comparison with that in wheat, *Cereal Chem.* **30:** 441.

HINTON, J. J. C. (1959), The distribution of ash in the wheat kernel, *Cereal Chem.* **36:** 19.

HINTON, J. J. C., PEERS, F. G. and SHAW, B. (1953), The B-vitamins in wheat: the unique aleurone layer, *Nature, Lond.* **172:** 993.

HINTON, J. J. C. and SHAW, B. (1953), The distribution of nicotinic acid in the rice grain, *Brit. J. Nutrit.* **8:** 65.

HODSON, A. Z. and NORRIS, L. C. (1939), A fluorometric method for determining the riboflavin content of foodstuffs, *J. Biol. Chem.* **131:** 621.

HORDER, T., DODDS, E. C. and MORAN, T. (1954), *Bread*, Constable, London.

HUBBARD, J. E., HALL, H. H. and EARLE, F. R. (1950), Composition of the component parts of the sorghum kernel, *Cereal Chem.* **27:** 415.

HUGHES, B. P. (1960), Amino acids, in McCANCE, R. A. and WIDDOWSON, E. M. (1960), *The Composition of Foods*, Med. Res. Coun., Spec. Rep. Ser. No. 297, H.M.S.O.

IHDE, A. J. and SCHUETTE, H. A. (1941), Thiamine, nicotinic acid, riboflavin and pantothenic acid in rye and its milled products, *J. Nutrit.* **22:** 527.

JONES, R. W., BABCOCK, G. E., TAYLOR, N. W. and SENTI, F. R. (1961), Molecular weights of wheat gluten fractions, *Arch. Biochem. Biophys.* **94:** 483.

JULIANO, B. O., BAUTISTA, G. M., LUGAY, J. C. and REYES, A. C. (1964), Rice quality studies on physicochemical properties of rice, *J. Agric. Fd. Chem.* **12:** 131.

KENT, N. L. (1946), Mineral constituents of wheat, flour and bran, in: BOOTH, R. G. *et al.* (1946).

KENT-JONES, D. W. and AMOS, A. J. (1957), *Modern Cereal Chemistry*, 5th edition, Northern Publ. Co. Ltd., Liverpool.

KIK, M. C. and LANDINGHAM, F. B. VAN (1943), Riboflavin in products of commercial rice milling and thiamin and riboflavin in rice varieties, *Cereal Chem.* **20:** 563.

KOCH, R. B., GEDDES, W. F. and SMITH, F. (1951), The carbohydrates of *Gramineae*. I. The sugars of the flour of wheat (*Triticum vulgare*), *Cereal Chem.* **28:** 424.

McCANCE, R. A., WIDDOWSON, E. M., MORAN, T., PRINGLE, W. J. S. and MACRAE, T. F. (1945), The chemical composition of wheat and rye and of flours derived therefrom, *Biochem. J.* **39:** 213.

McDERMOTT, E. E. and PACE, J. (1959), Extraction of the protein from wheaten flour in the form of soluble derivatives, *Nature, Lond.* **184:** 546.

McDERMOTT, E. E. and PACE, J. (1960), Comparison of the amino-acid composition of the proteins in flour and endosperm from different types of wheat, *J. Sci. Fd. Agric.* **11:** 109.

McLEOD, A. M. and WHITE, H. B. (1961), Lipide metabolism in germinating barley, *J. Inst. Brewing* **67:** 182.

McNEILL, J. R. (1952), Unpublished data. Quoted by MATZ, S. A. (1959).

MATZ, S. A. (1959) (Ed.), *The Chemistry and Technology of Cereals as Food and Feed*, Avi. Publ. Co., Westport, Conn., U.S.A.

MEYER, J. H., GRUNERT, R. R., GRUMMER, R. H., PHILI IPS, P. H. and BOH-STEDT, G. (1950), Sodium, potassium and chlorine content of feeding stuffs, *J. Animal Sci.* **9**: 153.

MICKUS, R. R. (1959), Rice (*Oryza sativa*), *Cereal Sci. Today* **4**: 138.

MOIR, H. C. (1946), The chemical composition and nutritive value of oats and oatmeal, in: BACHARACH, A. L. and RENDLE, T. (1946), *The Nation's Food*, Soc. Chem. Ind., London.

MORAN, T. (1959), Nutritional significance of recent work on wheat, flour and bread, *Nutr. Abs. Revs.* **29**: 1.

MUIR, quoted by GRIST, D. H. (1959), p. 331.

NELSON, quoted by WATSON, S. J. (1953).

NELSON, G. H., TALLEY, L. E. and ARONOVSKY, S. I. (1950), Chemical composition of grain and seed hulls, nut shells and fruit pits, *Trans. Amer. Assoc. Cereal Chem.* **8**: 58.

NELSON, J. H., GLASS, R. L. and GEDDES, W. F. (1963), The triglycerides and fatty acids of wheat, *Cereal Chem.* **40**: 343.

NIELSEN, H. C., BABCOCK, G. E. and SENTI, F. R. (1962), Molecular weight studies on glutenin before and after disulphide bond splitting, *Arch. Biochem. Biophys.* **96**: 252.

OKAMURA, T., HAYAKAWA, K. and KIMATA, N. (1961), Inorganic components of foods. II. The content of fluorine in rice produced in Aichi prefecture, *Aichi Gakugei Daigaku Kenkyu Hokoku* **10**: 159.

OSBORNE, T. B. (1907), The proteins of the wheat kernel, *Carnegie Inst. Wash., Pub.* No. 84.

PEDRERO, P. S. (1962), A polarographic method for the determination of the manganese content of cereals grown in Spain, *Anal. Real. Acad. Farm.* **28**: 61.

PENCE, J. W., ELDER, A. H. and MECHAM, D. K. (1951), Some effects of soluble flour components on baking behaviour, *Cereal Chem.* **28**: 94.

PENCE, J. W. and NIMMO, C. C. (1964), New knowledge of wheat proteins, *Bakers Digest* **38** (1): 38.

PENCE, J. W., NIMMO, C. C. and HEPBURN, F. N. (1964), Proteins, in: HLYNKA, I. (1964) (Ed.), *Wheat: Chemistry and Technology*, Amer. Assoc. Cereal Chem., St. Paul, Minn., U.S.A.

PENCE, J. W., WEINSTEIN, N. E. and MECHAM, D. K. (1954), The albumin and globulin contents of wheat flour and their relationship to protein quality, *Cereal Chem.* **31**: 303.

SCHMORL, K. (1933), Die Mineralstoffverteilung im Getreidekorn, *Z. Ges. Getreide Muhlen. Bäkereiw.* **20**: 300.

SCHOPMEYER, H. H. (1962), Rye and rye milling, *Cereal Sci. Today* **7**: 138.

SCHULTZ, A. S., ATKIN, L. and FREY, C. N. (1941), A preliminary survey of the vitamin B_1 content of American cereals, *Cereal Chem.* **18**: 106.

SEBRELL, W. JR. and HARRIS, R. S. (1954), *The Vitamins—Chemistry, Physiology, Pathology*, Vol. 3, Academic Press, New York.

SHAW, B. (1953), Unpublished data.

SHERMAN, H. C. (1941), *Chemistry of Food and Nutrition*, 6th edition, Macmillan, New York.

SHOLLENBERGER, J. H. and JAEGER, C. M. (1943), Corn—its products and uses, *Northern Regional Research Laboratory, Peoria, Ill., U.S.A., Bulletin.*

STEVENS, D. J., McDERMOTT, E. E. and PACE, J. (1963), Isolation of endosperm protein and aleurone cell contents from wheat, and determination of their amino-acid composition, *J. Sci. Fd. Agric.* **14**: 284.

THOMAS, J. M., BINA, A. F. and BROWN, E. B. (1942), The nicotinic acid content of cereal products, *Cereal Chem.* **19**: 173.

VAISEY, M. and UNRAU, A. M. (1964), Flour composition, chemical constituents of flour from cytologically synthesized and natural cereal species, *J. Agric. Fd. Chem.* **12**: 84.

WATSON, S. J. (1953), The quality of cereals and their industrial uses. The uses of barley other than malting, *Chem. & Ind.*, p. 95.

WATT, B. K. and MERRILL, A. L. (1950), Composition of Foods. U.S. Dept. Agric., *Agric. Handbook* 8.

WILLIAMS, V. R., KNOX, W. C. and FIEGER, E. A. (1943), A study of some of the vitamin B-complex factors in rice and its milled products, *Cereal Chem.* **20**: 560.

WOYCHIK, J. H., BOUNDY, J. A. and DIMLER, R. J. (1961), Starch gel electrophoresis of wheat gluten proteins with concentrated urea, *Arch. Biochem. Biophys.* **94**: 477.

FURTHER READING

HLYNKA, I. (Ed.), *Wheat: Chemistry and Technology*, Amer. Assoc. Cereal Chem., St. Paul, Minn., U.S.A., 1964.

McDERMOTT, E. E. and PACE, J., The content of amino-acids in white flour and bread, *Brit. J. Nutr.* **11**: 446, 1957.

MORAN, T., Nutrients in wheat endosperm, *Nature, Lond.* **155**: 205, 1945.

MORAN, T., Nutrients in British-grown and imported wheat, *Nature, Lond.* **157**: 643, 1946.

MORRISON, F. B., *Feeds and Feeding*, 20th edition, Morrison Publ. Co., Ithaca, N.Y., 1947.

PACE, J., Recent work on the proteins of wheat flour, *Recent Advances in Processing Cereals*, pp. 99–112. Monograph 16, Soc. Chem. Ind., London, 1962.

CHAPTER 4

WHEATS OF THE WORLD

DISTRIBUTION

Wheat is grown throughout the world, from the borders of the arctic to near the equator, although the crop is most successful between latitudes 30 and 60 North and between 27 and 40 South. In altitude, it ranges from sea level to 10,000 ft in Kenya and 15,000 ft in Tibet. Cultivated varieties, which are of widely differing pedigree and are grown under varied conditions of soil and climate, show wide variations in characteristics.

Soil

Wheat grows best on heavy loam and clay, although it makes a satisfactory crop on lighter land. The crop repays heavy nitrogenous manuring.

Climate

Wheat flourishes in subtropical, warm temperate and cool temperate climates. An annual rainfall of 9–30 in., falling more in spring than in summer, suits it best. The mean summer temperatures should be 56°F or more.

SPRING AND WINTER WHEAT

The seed is sown in late autumn (winter wheat) or in spring (spring wheat).

Winter wheat

This can be sown in places, e.g. north-western Europe, where excessive freezing of the soil does not occur. The grain germinates in the autumn and grows slowly until the spring. Frost would affect the young plants adversely, but a covering of snow protects them and promotes tillering.

Spring wheat

In countries such as the Canadian prairies and the steppes of Russia that experience winters too severe for winter sowing, wheat is sown as early as possible in the spring, so that the crop may be harvested before the first frosts of autumn.

The climatic features in countries where spring wheat is grown —maximum rainfall in spring and early summer, and maximum temperature in mid- and late summer—favour production of rapidly maturing grain with endosperm of vitreous texture and high protein content, suitable for breadmaking. The area of production of spring wheat is being extended progressively northwards by the use of new varieties bred for their quick-ripening properties.

Winter wheat, grown in a climate of relatively even temperature and rainfall, matures more slowly, producing a crop of higher yield and lower protein content, better suited for biscuit and cake making than for bread.

RACES AND SPECIES OF WHEAT

The known species and varieties of the genus *Triticum* (wheat) are said to number over 30,000; all can be grouped into three distinct races which have been derived from separate original ancestors, and which differ in chromosome numbers. One classification of the races, with the probable wild types, chromosome number (2n) and the cultivated forms, is shown in Table 27.

Einkorn, emmer and spelt are husked wheats, i.e. the lemma and palea form a husk which remains attached to the kernel after threshing. Emmer was used for human food in prehistoric times (cf. p. 114); there is archaeological evidence that it was grown about 5000 B.C. in Iraq.

TABLE 27
A CLASSIFICATION OF WILD WHEAT TYPES

Race	Wild type	2n	Cultivated forms	
			Species name	Common name
Small spelt	*T. aegilopoides*	14	*T. monococcum*	Einkorn
Emmer	*T. dicoccoides*	28	*T. dicoccum*	Emmer
			T. durum	Macaroni wheat (durum)
			T. polonicum	Polish
			T. turgidum	Rivet, Cone
Large spelt ⎫	probably	42	*T. vulgare*	Bread wheat
Dinkel ⎭	*T. dicoccoides* ×		*T. spelta*	Dinkel, Spelt
	Aegilops ovata or		*T. compactum*	Club
	Ae. cylindrica		*T. sphaerococcum*	Indian Dwarf

The principal wheats of commerce are varieties of the species *T. durum*, *T. vulgare* and *T. compactum*.

WHEAT TYPES

In a general way, wheats are classified according to (1) the texture of the endosperm, because this characteristic of the grain is connected with the way the grain breaks down in milling, and (2) the protein content, because the properties of the flour and its suitability for different purposes are related to this characteristic.

Vitreous and mealy wheats

The endosperm texture may be vitreous (steely, flinty, glassy, horny) or mealy (starchy, chalky). Samples may be entirely vitreous or entirely mealy, or may consist of a mixture of vitreous and mealy grains, with one type predominating. Individual grains are generally completely vitreous or completely mealy, but grains which are partly mealy and partly vitreous ("piebald grains") are frequently encountered. The specific gravity of vitreous grains is generally higher than that of mealy grains: 1·422 for vitreous and 1·405 for mealy (Bailey, 1916).

The vitreous or mealy character is hereditary, but also affected by environment. Thus, *T. aegilopoides*, *T. dicoccoides*, *T. mono-*

coccum and *T. durum* are species with vitreous kernels, whereas *T. turgidum* and many varieties of *T. compactum* and *T. vulgare* are mealy (Percival, 1921). However, the vitreous/mealy character may be modified by the cultural conditions. Thus, the percentage of vitreous kernels in a white English variety (Holdfast) was 90 in a sample grown on clay, but only 20 in a sample grown on gravel in a poor harvest (Greer, 1949). Mealiness is favoured by heavy rainfall, light sandy soils, and crowded planting, and is more dependent on these conditions than on the type of grain grown. Vitreousness can be induced by nitrogenous manuring or commercial fertilizing and is positively correlated with high protein content; mealiness is positively correlated with high grain-yielding capacity.

Vitreous kernels are translucent and appear bright against a strong light, whereas mealy kernels are opaque, and appear dark under similar circumstances.

The opacity of mealy kernels is an optical effect due to minute vacuous or air-filled fissures between and perhaps within the endosperm cells. The fissures form internal reflecting surfaces, preventing light transmission and giving the endosperm a white appearance. Such fissures are absent from vitreous endosperm, in which the cells are completely filled with starch–protein matrix.

The development of mealiness seems to be connected with maturation, since immature grains of all wheat types are vitreous, and vitreous grains are found on plants that grow and ripen quickly: spring wheats, and those growing in dry continental climates. Mealy grains are characteristic of varieties that grow slowly and take a long time to ripen.

Vitreous kernels sometimes acquire a mealy appearance after being conditioned in various ways, e.g. by repeated damping and drying or by warm conditioning (cf. p. 108). The proportion of vitreous kernels in the sample is a characteristic used in the U.S. wheat grading system (see p. 73).

Wheat types may also be classified as hard or soft, and as strong or weak (see p. 68). Vitreous grains tend to be hard and strong, mealy grains to be soft and weak, but the association is not invariable.

Hard and soft wheats

"Hardness" and "softness" are milling characteristics relating to the way the endosperm breaks down. Greer and Hinton (1950) observed that, if the cut surface of hard wheat is lightly and uniformly wetted and allowed to dry, a pattern of cracks appears, following the lines of the endosperm cell boundaries. When soft wheat is treated similarly, the pattern of cracks produced bears no relationship to the cellular structure of the endosperm (which resembles that in hard wheat) but passes indiscriminately through the cells. This phenomenon suggests a pattern of areas of mechanical strength and weakness in hard wheat, but fairly uniform mechanical weakness in soft wheat.

Hard wheats yield coarse, gritty flour, free-flowing and easily sifted, consisting of regular-shaped particles, which are mostly whole endosperm cells; soft wheat gives very fine flour consisting of irregular-shaped fragments of endosperm cells (including a proportion of quite small cellular fragments and free starch granules), with some flattened particles, which adhere together, sift with difficulty, and tend to close the apertures of sieves (cf. p. 142).

According to Berg (1947), hardness is a milling characteristic that is transmitted by breeding, and is inherited in Mendelian fashion. The endosperm of hard wheats may be flinty or mealy in appearance, but its breakdown is always typical of a hard wheat. It is probable that the strength of the protein bonds determines the nature of the breakdown.

Hardness affects the ease of detachment of the endosperm from the bran. In hard wheats the endosperm cells come away more cleanly and remain more intact, whereas in soft wheats the peripheral endosperm cells tend to fragment, part coming away while part is left attached to the bran.

The granularity of flour gives a measure of the relative hardness of wheats, the proportion of the flour passing through a fine flour silk decreasing with increasing hardness. Berg (1947) found that the percentage flour yield through a No. 15 silk (aperture width: 0·09 mm) was related to wheat variety, and not to baking quality. Greer (1949) found that the percentage of the total flour passing through a No. 16 silk was 49–56% for four related varieties of hard English wheats, viz. Yeoman and three varieties related to Yeoman, whereas it was 63–71% for ten varieties of soft English wheat unrelated to Yeoman. These comparisons relate to tests carried out under standard conditions; the ease of sifting, as distinct from the granularity of the flour, is affected by other factors, besides hardness of the endosperm (cf. p. 103).

The principal wheats of the world are arranged according to their degrees of hardness as follows:

Extra hard	Durum, some Algerian, Indian
Hard	Manitoba, American HRS
Medium	Plate, Australian, American HRW
Soft	European, American SRW

Strong and weak wheats

Strength is a characteristic of wheat relating to its baking strength, viz. the ability of the flour to produce bread of large loaf volume and good crumb texture (cf. pp. 147 and 179). Wheats with this characteristic generally have a high protein content and are called "strong", whereas those from which only a small loaf, with coarse open crumb structure, can be made, and which are characterized by low protein content, are called "weak". The flour from weak wheats is ideal for biscuits (cookies) and cakes, although unsuitable for breadmaking. Other features of flour from strong wheats are its ability to carry a proportion of weak flour, i.e. for the loaf to maintain its large volume and good crumb structure even when the flour is blended with a proportion of weak flour, and its ability to absorb and retain a large quantity of water.

The main types of wheat are classified according to their baking strength as follows:

Strong   Manitoba, American HRS, Russian spring
Medium American HRW, Plate, S.E. European
Weak   N.W. European, American SRW, Australian

Hardness (milling character) and strength (baking character) are not closely linked genetically, but appear to segregate separately (Berg, 1947). Hence, it should be possible, through breeding, to combine good milling quality with, for example, the type of gluten associated with soft wheats. The Swedish variety Eroica is hard but possesses no particular baking strength; Fylgia, of high breadmaking quality, is derived from Kolben, an excellent ~~milling~~ baking wheat, but possesses the milling character of soft wheats.

PROTEIN CONTENT AND MILLING QUALITY

Protein content is not a factor determining milling quality, except in so far as the protein content tends to be higher in vitreous than in mealy wheats, and vitreousness is often associated with hardness and good milling quality. Humphries (1923) found that the milling character of Red Fife wheat grown in England remained constant over a 9–14% range of protein content. Samples of the English soft wheat varieties Cappelle Desprez or Hybrid 46 may have high protein content and a large proportion of vitreous grains and yet mill as soft wheats; on the other hand, a low protein, predominantly mealy-grained sample of the hard variety Maris Widgeon will mill as a hard wheat. Protein content is, however, a most important characteristic in determining baking quality (see pp. 68, 71, 147, 158).

GRAIN SIZE AND SHAPE

The maximum yield of flour, obtainable from wheat in milling, is ultimately dependent on the endosperm content, and the latter is affected by the size and shape of the grains, and by the thickness of the bran.

The bushel weight measurement (test wt. per bu in the U.S.A.; natural wt. in Europe) estimates the weight of a fixed volume of grain and gives a rough indication of kernel size and shape. Wheats of high bu wt. are usually considered to mill the more readily and to yield more flour. However, these measurements can be misleading, as weak mealy wheats often have high bu wt.

Shellenberger (1961) found that the volumetric bran content is lower in large than in small grains, viz. 14·6% and 14·1%, respectively, from samples of the same type of wheat, showing the economic importance of large kernel size.

WORLD WHEATS

Under normal conditions, the British flour-miller purchases wheat from all over the world, and can produce flour comparing favourably in quality and price with flour milled elsewhere.

Although the miller uses a great diversity of wheats, he ensures that his flours are of regular quality by skilful blending of the different types so that particular properties lacking in one component of the grist may be provided by another.

Time of supply

Times of sowing and of harvesting the wheat crop in the various growing countries are naturally dependent on local climatic conditions; wheat is being harvested in some country in every month of the year. However, the storage facilities in most wheat-growing countries are adequate to permit the best part of a year's harvest being stored; thus, the British miller can buy wheat from any exporting country at almost any time of the year.

The times of harvest for the principal wheat-growing countries are shown in Table 28.

CHARACTERISTICS OF WORLD WHEAT TYPES

America

The principal wheat-producing countries of the American continent are Canada, the U.S.A. and Argentina.

Canada

Over 95% of the Canadian wheat crop is spring sown, and the grains are hard, red-skinned, and with vitreous endosperm and a high content of protein which makes gluten of good extensibility and elasticity, ideally suited for breadmaking. Canadian wheat, which is called "Manitoba", is a strong wheat, and of excellent milling quality, although not so hard as durum. It is grown in the provinces of Manitoba, Saskatchewan and Alberta.

TABLE 28
TIMES OF WHEAT HARVEST

Country	Harvest	Arrival of new crop
Canada	Aug., Sept.	Nov.
Australia ⎫ Argentina ⎭	Dec., Jan.	Feb., Mar.
India	Feb.–Apr.	June, July
China	May	June, July
Italy	June	June, July
France	June, July	July, Aug.
U.S.A.	May–Aug.	Aug.–Oct.
Russia	July–Sept.	Aug.–Oct.
England	July, Aug.	July–Sept.

* From Kent-Jones and Amos (1957).

The varieties Thatcher and Selkirk occupied respectively 43% and 28% of the total Canadian wheat acreage in 1957. They were bred to replace Marquis, a good milling variety which was not resistant to rust but which usually escaped damage because it ripened early. Thatcher resulted from the double cross (Iumillo × Marquis) × (Kanred × Marquis). It outyields Marquis, and it was resistant to stem rust when released, but proved susceptible to race 15B, to which Selkirk is resistant. Both varieties are resistant to loose smut, and Selkirk is also resistant to bunt, and moderately resistant to leaf rust. Other Canadian varieties are still graded with reference to Marquis, although the latter is seldom grown commercially today.

The protein content of Manitoba wheat varies 9–18%, and the yield is now about 9·5 cwt/ac, the yield and the protein content tending to be related inversely. The moisture content at harvest is usually 11–13%.

Manitoba Red Spring wheat is classified as Atlantic or Pacific (Vancouver), according to port of shipment, and is graded, according to bu wt., percentage content of foreign material and seeds, and percentage content of wheat of other classes and varieties not equal to Marquis, into five grades: Nos. 1, 2, 3 and 4 Manitoba Northern, and No. 5 Wheat. For inclusion in these grades the m.c. must not exceed 14·5%. Wheat with 14·6–17% m.c. is graded "tough"; with over 17% "damp". Wheat that does not qualify for a statutory or commercial grade is described as "sample" grade.

Early frosts may reduce the yield of grain and lower its milling quality by increasing the proportion of small shrivelled grains with low endosperm content, and adversely affect baking quality, because the milled flour is of high maltose content and produces a flowy dough.

A small amount of winter wheat (autumn sown) is grown in southern Alberta (hard wheat), and in British Columbia and western Ontario (soft wheat).

The wheat known as Canadian Eastern is low in protein content (about 9%); it is suitable for high ratio cake flours (cf. p. 150) and for biscuits, when mixed with more extensible wheats. Being low in diastatic power, it is also suitable for sausage rusk which requires flour of low maltose content and high absorbency.

United States

Five principal types of wheat are grown in the U.S.A.: their names, the proportion that each contributed to the total crop (average over 10 years, 1952–3 to 1961–2), their average bu wt. and protein contents, are shown in Table 29.

Hard Red Spring wheat is grown in Minnesota, North Dakota, Montana and South Dakota. The milling quality is only slightly inferior to that of Manitoba, and the protein content is comparable. HRS is graded according to content of dark, hard and

TABLE 29
CHARACTERISTICS OF U.S. WHEAT TYPES *

Type	Proportion (%)	Bu wt. (lb)	Protein Range (%)	Protein Average (%)
Hard Red Winter (HRW)	52	62–64	9·6–14·8	11·7
Soft Red Winter (SRW)	16	60–64	8·8–11·1	10·3
Hard Red Spring (HRS)	15·5	63–64	10·5–12·8	12·4
White, hard	14·5	61–63	10·2–11·4	10·8
White, soft			8–10	9·0
Durum	2	63–64	—	13·0

* Data from Dahl (1962), Johnson (1962), Shellenberger (1961).

vitreous kernels into Dark Northern Spring (75% or more), Northern Spring (25–75%) and Red Spring (less than 25%). HRS is used for quality yeasted breads and rolls.

Hard Red Winter wheat is grown in Texas, Oklahoma, Kansas, Colorado, Nebraska, Montana, South Dakota and Minnesota. Exports of HRW from the Gulf of Mexico are generally higher in protein content (14%) than the remainder ($11\frac{1}{2}$–$12\frac{1}{2}$%), and slightly stronger in baking quality. HRW is used for making yeasted bread and hard rolls. Dark Hard Winter must contain 75% or more of dark, hard and vitreous kernels; Hard Winter 40–75%; Yellow Hard Winter less than 40%.

White. White wheat includes hard and soft types; Hard White must contain 75% or more of hard kernels. Soft White is also known as Pacific White in Britain. White wheat is grown in the west coast States, and in Michigan and New York. The flour from white wheat is unsuitable for breadmaking, but is ideal for crackers (biscuits), cakes and pastries (cookies). The soft white is similar to Canadian Eastern white wheat, but slightly stronger and higher in protein content. The diastatic power tends to be low.

Soft Red Winter wheat is grown in Missouri, Illinois, Ohio, Indiana and New York States. That grown east of the Great Plains is called Red Winter; the remainder is Western Red. SRW mills well but is a weak wheat, low in protein content.

The flour is used for biscuits and crackers, cakes and pastries (cookies).

Durum wheat is grown in North and South Dakota. It is milled to provide semolina (p. 116) for making macaroni, spaghetti, noodles, etc. (cf. p. 193). It is a very hard wheat, but the flour is unsuitable for breadmaking. Hard Amber Durum has 75% or more of hard and vitreous kernels of amber colour; Amber Durum has 60–75%; Durum has less than 60%.

U.S. wheat grading

The main classes of wheats are graded according to test wt. per bu, content of damaged and shrunken kernels, foreign material and wheat of other classes, as shown in Table 30.

Certain special classes are recognized. "Heavy" wheat is HRS wheat of grades 1, 2 or 3 with test wt. of 60 lb or more per bu, or wheat of other classes, of grades 1, 2 or 3, with test wt. of 62 lb or more per bu. The word "tough" is added if the m.c. exceeds 13·5% (or is between 14·5% and 16% for durum). Parcels which do not fall within the limits of grades 1–5, or are of low quality in certain respects, and durum wheat of over 16% m.c., are designated "sample grade". "Smutty wheat" contains more than 14 smut balls per 250 g (cf. p. 85). "Garlicky wheat" contains 2 or more green garlic bulbils per kg. "Weevily wheat" is wheat infested with live weevils or other insects injurious to stored grain (cf. p. 130). "Treated wheat" has been scoured, limed, washed, sulphured or treated in such a manner that the true quality is not reflected by either the numerical grade or sample grade designation alone. (Abstracted from Official Grain Standards of the United States, effective 1 May 1964 *Federal Register*, 1 June 1964.

Price support

In addition to the grading system described above, a price support scheme is operated in the U.S.A., whereby the market price of grain is maintained by a system of government purchase

TABLE 30

GRADE CHARACTERISTICS OF U.S. WHEAT*

Grade	Minimum test wt. (lb/bu)		Maximum limits (%) of					Wheat of other classes	
	HRS	Other classes	Heat damaged kernels	Damaged kernels (total)	Foreign material	Shrunken and broken kernels	Defects (total)	Contrasting	Total
1	58	60	0·1	2	0·5	3	3	0·5	3
2	57	58	0·2	4	1·0	5	5	1	5
3	55	56	0·5	7	2·0	8	8	2	10
4	53	54	1·0	10	3·0	12	12	10	10
5	50	51	3·0	15	5·0	20	20	10	10

* Source: U.S. Dept. Agric. (1964).

and loans to growers according to the quality of their produce. Under earlier support programmes, premiums ranging from 1 to 12 cents per bu were paid for protein content from 11 to 17%, with additional premium for content over 17%. From the 1962 harvest, however, the price support scheme has been based on the results of the Zeleny sedimentation test (which is said to reflect both protein content and protein quality—but see p. 152). Premiums are provided for wheats with baking quality values of 40 or above, as measured by the sedimentation test (see Table 31).

TABLE 31
U.S. WHEAT PREMIUMS

Sedimentation test value	Premium cents/bu
40–44	3
45–49	6
50–54	10
55–59	14
60–64	19
65 and over	24

U.S. wheat surplus

Since 1952 the surplus of wheat left at the end of the season in the U.S.A. when the new crop is coming in (known as the "carry over") has increased greatly. By 1962, the carry over amounted to 1360 million bushels, or considerably more than one entire U.S. wheat harvest (1150 million bushels in 1962). The composition of the carry over in 1952 and in 1962 is shown in Table 32.

The bulk of the carry-over wheat is now HRW, although this type constituted only 52% of the production in 1962. Moreover, much of the accumulated HRW wheat is of low protein content. The factors leading to this situation include:

1. The unusually low protein content of the 1961 crop.
2. Selection of the better parcels of HRW (of higher protein content) for domestic use or export.

3. The trend towards replacing HRS by HRW, e.g. in Montana, South Dakota, Wyoming, Colorado and Idaho, and towards replacing SRW by HRW, e.g. in Missouri, Kentucky, Tennessee and Illinois. Replacement of HRS or SRW by HRW has occurred because of: (a) the development of higher yielding and more adaptable varieties of HRW; (b) the small and relatively constant differentials in support price between types and qualities of wheat.

TABLE 32
COMPOSITION OF U.S. WHEAT CARRY OVER *
(MILLION BU)

Year	HRW	SRW	HRS	Durum	White	Total
1952	97	16	117	15	11	256
1962	1127	30	171	2	30	1360
Added in 10 years	1030	14	54	—13	19	1104

* Data from Dahl (1962).

The HRW grown in areas that formerly carried HRS is inferior in breadmaking quality to the wheat it has replaced, e.g. HRS on the eastern edge of the Great Plains area.

Argentina

Argentina is the main producer of bread wheat in South America. The wheat is classified as Hard Red Winter, and is known as "Plate" wheat. The grain is hard, red and semi-vitreous, small, thin and elongated in shape. In quality, the wheat is strong, with about 12% protein content, but the gluten has limited extensibility, and the wheat is low in diastatic power, and suitable only as a filler wheat in breadmaking grists.

The types of Plate wheat are named after the ports of shipment: Rosafé, grown in the north, around Rosario and Sante Fé, is shipped from Rosario; Baril, grown in the central area, from Buenos Aires; Barusso, grown in the south, from Bahia Blanca. Barusso is a softer wheat, with lower protein content, than the other types.

Australia

Wheat is grown in the relatively high rainfall areas of New South Wales, Victoria, South Australia, Western Australia and Queensland. There are two classes: hard and soft wheats. The hard type is of medium strength, and is suitable as a filler in breadmaking grists. The soft type is weak, and the flour is good for biscuits and pastry production. Australian wheat exports arc mostly of the soft type. The grains are large, dry, brittle and slightly elongated, with thin bran of white or yellow colour. The m.c. is low, seldom above 11%, and the grain is millable up to 16% m.c.

Recent attempts to raise the protein content by undersowing with clover have already produced promising results.

Europe

U.S.S.R. Good-quality Hard Red Spring, Hard Red Winter and Durum wheats are produced. The grains of the HRS and HRW are small, red, hard and horny, and the wheat may contain frosted grains and be damaged by wheat bug (cf. p. 86). The big climatic variation within the U.S.S.R. gives rise to a wide range of wheat quality from very strong to medium strong. Average Russian wheat is weaker than Manitoba in baking strength and is suitable as a filler wheat. The protein content averages 12%.

Britain

Winter and spring types are grown, with winter predominating. Bran colour may be red or white, but most varieties are soft wheats of little use for breadmaking, although suitable for biscuits, self-raising flour and pastry. Much of the grain is harvested at relatively high m.c. (16–20%) and needs drying. In 1962, the variety Cappelle Desprez made up about 70% of the winter wheat, Jufy about 60% of the spring. These varieties will eventually be replaced by higher yielding types.

The protein content may range from 8 to 13%, according to

locality, within one season. At any one locality, the average range of protein contents between varieties would be about 2%, the higher yielding varieties having the lower protein contents.

There is no official grain grading scheme in Britain. Wheat is described as "millable" or "non-millable", but the large variation in quality within the former class is not recognized by any official system of price support.

The National Institute of Agricultural Botany (headquarters at Cambridge) undertakes thorough examination of new wheat varieties and issues recommendations to British farmers. The milling and baking quality of contemporary varieties is assessed by the Research Association of British Flour-Millers: a section of their current *Classification of Home-grown Wheat Varieties* (by Greer and Stewart) is shown in Fig. 10.

About 6% of British wheat is used for seed; the remainder is shared almost equally between flour-milling and animal feeding requirements.

Other western European wheat

Wheat from France, Germany, Belgium and Baltic states is similar to British, but is harvested somewhat drier (15–17% m.c.) and is a little stronger.

Grain production in France has risen from 8 to 11 million tons per annum, despite a 10% shrinkage in acreage, during the past 25 years, because yield per acre has increased from 12 to 20 cwt. Total storage capacity has increased from 2·5 to 7 million tons (Nuret, 1963).

South-eastern Europe

Wheat from Bulgaria and Roumania is somewhat harder and stronger than that from the west and can be used as a filler in breadmaking grists.

Variety	Parentage	Season	Grain Colour	Grain Texture	Milling Value*	Bread-making Value*	Biscuit Value*	Water Absorp-tion*
Alex	Red Marvel × Garnet × Bersee	Spring	Red	Hard	2	2	2	1
Als	Alsen × Alsen	Winter	Red	Soft	4	3	3	–
Atle	Extra Kolben × Saxo	Spring	Red	Hard	1	1	3	2
Atson	(Marquis × Hatif Inversable) × WS 8473	Spring	Red	Hard	1	1	4	2
Ayr Challenge	Hybrid 46 × Holdfast	Winter	Red	Soft	2	3	3	3
Banco	Bankuter 178 and Swedish varieties	Winter	Red	Hard	2	2	4	2
Belle (see Elite Lepeuple)								
Benign	Holdfast × Benoist 40	Winter	Red	Hard	1	2	4	2
Bersee	Hybride des Alliés × Vilmorin 23	Winter	Red	Soft	4	4	1	3
Cappelle Desprez	Hybride du Joncquois × Vilmorin 27	Winter	Red	Soft	2	3	3	3
Carpo	Lin Calel × Bastard II × Jabo	Spring	Red	Soft	4	3	4	2
Cash & Courage	Demeter × Jubilegem	Winter	Red	Soft	3	4	2	3
Champlein	Yga Blondeau × Tadepi	Winter	Red	Soft	2	3	3	3
Dominator	Jubilegem × Atle	Winter	Red	Hard	1	2	3	2
Druchamp	Vilmorin 27 × Flèche d'Or	Winter	White	Soft	3	3	3	–
Eclipse	Yeoman × Little Joss	Winter	Red	Soft	3	3	3	3
Elite Lepeuple (Belle)	Bellevue × Bersee	Winter	Red	Hard	1	2	3	2
Falco	Minister × Lovink	Winter	White	Soft	2	4	1	4
Felix	(Tassilo × Carsten) × (Carsten × Mar-quillo)	Winter	Red	Soft	3	4	4	3
Flamingo	(Tassilo × Kron) × Heines IV	Winter	White	Soft	4	4	1	4
Fylgia	Extra Kolben II × Aurora	Spring	Red	Soft	3	2	3	2
Fylgia II	Extra Kolben II × Aurora	Spring	Red	Hard	2	2	3	1
Glasnevin Rosa	Desprez 80 × (Iron × Squarehead's Master)	Winter	White	Soft	4	4	1	3
Heines VII	Hybride à Courte Paille × Svalöf Kron	Winter	Red	Soft	3	4	2	3
Hesbignon	(Hybride du Joncquois × Prof. Delos) × (Bastard II × Prof. Delos)	Winter	Red	Hard	2	3	2	1
Holdfast	Yeoman × White Fife	Winter	White	Hard	1	1	4	1
Hybrid 46	Benoist 40 × other hybrids	Winter	Red	Soft	3	4	2	3
Jubilegem	Vilmorin 23 × Iron III	Winter	Red	Soft	3	4	2	–
Jufy I	Jubilegem × Fylgia	Spring	Red	Soft	4	3	3	3
Jufy II	Jubilegem × Fylgia	Spring	Red	Soft	3	4	3	3
Juliana	Wilhelmina × Essex Smooth Chaff	Winter	White	Soft	3	3	1	3
Karn	(Marquis × Hatif Inversable) × (Extra Kolben × Dutch land variety)	Spring	Red	Hard	1	1	4	–
Karn II	(Marquis × Hatif Inversable) × (Extra Kolben × Dutch land variety)	Spring	Red	Hard	1	1	4	1
Koga II	(Heines Kolben × Garnet) × (Heines Kolben × Raeckes Whitechaff)	Spring	Red	Hard	3	2	4	1
Leda	Jubilegem × Zanda	Winter	Red	Soft	3	3	2	3
Little Joss	Squarehead's Master × Ghirka	Winter	Red	Soft	3	4	1	4
Maître Pierre	Druchamp × Yga Blondeau	Winter	Red	Hard	2	2	4	1
Marne Desprez	Cappelle × (H80 × PLM)	Winter	Red	Soft	2	2	4	2
Marsters 57	Yeoman × Hybrid 46	Winter	Red	Soft	2	4	2	4
Masterpiece	Holdfast × Squarehead's Master	Winter	Red	Hard	1	1	4	1
Miana	Institut Agronomique × Bersee	Winter	Red	Soft	3	4	1	3
Milfast	Atle × Holdfast	Winter	Red	Hard	1	1	4	1
Minister	Benoist 40 × Prof. Delos	Winter	White	Soft	2	3	1	4
N. 59	Single plant selection	Winter	Red	Hard	1	3	3	1
Opal	Triesdorfer Ruf × Garnet × Heines Kolben × Koga × Rümkes Erli	Spring	Red	Hard	2	2	4	2
Peko	Peragis × Heines Kolben	Spring	Red	Soft	4	4	1	4
Petit Quin Quin	(Vilmorin 23 × Institut Agronomique) × Providence	Winter	Red	Soft	3	3	2	–
Phoebus	Jubilegem × Fylgia	Spring	Red	Soft	4	4	1	3
Pia	Vilmorin 27 × Hybride à Courte Paille	Winter	Red	Hard	1	3	3	2
Pilot	Little Tich × Swedish Iron	Winter	Red	Hard	1	4	4	–
Prestige	[(Bohemien Wheat × Rye) × Oro] × (Dômes × Garnet)	Winter	Red	Hard	2	2	3	2
Professeur Marchal	(Desprez 80 × Prof. Delos) × (Bastard II × Prof. Delos)	Winter	Red	Hard	2	4	2	2
Progress	A23/28 × Extra Kolben	Spring	Red	Hard	1	1	4	1
Redmace	Single ear selection	Winter	Red	Soft	2	3	3	3
Redman	Yeoman × Squarehead's Master	Winter	Red	Hard	2	1	3	2
Ring	Karn II × Pondus	Spring	Red	Hard	1	1	4	1
Ritchie	Atle × Garnet Hybrid	Spring	Red	Hard	2	2	3	2
Rivet	Selection from English land variety	Winter	Red	Hard	3	4	4	1

REFERENCES

BAILEY, C. H. (1916), The relation of certain physical characteristics of the wheat kernel to milling quality, *J. Agric. Sci.* **7**: 432.

BERG, S. O. (1947), Is the degree of grittiness of wheat flour mainly a varietal character?, *Cereal Chem.* **24**: 274.

DAHL, R. P. (1962), Classes of wheat and the surplus problem, *Northwestern Miller* **266** (8): 30.

GREER, E. N. (1949), A milling character of home-grown wheat, *J. Agric. Sci.* **39**: 125.

GREER, E. N. and HINTON, J. J. C. (1950), The two types of wheat endosperm, *Nature, Lond.* **165**: 746.

HUMPHRIES, A. E. (1923), *Report of the Home-grown Wheat Committee of the National Assoc. of British and Irish Millers*.

JOHNSON, J. A. (1962), *Wheat and Flour Quality*, Kansas State University, Mimeo Publ.

KENT-JONES, D. W. and AMOS, A. J. (1957), *Modern Cereal Chemistry*, 5th edition, Northern Publ. Co. Ltd., Liverpool.

NURET, H. (1963), Die Entwicklung der Müllerei und der Müllereimaschinen in Frankreich in den letzten 25 Jahren, *Die Mühle* **100**: 576.

PERCIVAL, J. (1921), *The Wheat Plant*, Duckworth, London.

SHELLENBERGER, J. A. (1961), World wide review of milling evaluation of wheats, *Assoc. Oper. Millers Bull.* 2620.

U.S. DEPARTMENT OF AGRICULTURE, Agricultural Marketing Service, Grain Division (1964), Official Grain Standards of the United States, Service and Regulatory Announcement AMS-177, *Federal Register*, June 1964.

FURTHER READING

Canadian Grain Handbook, Crop Year 1963–4, Canadian Wheat Board, Winnipeg.

FLOUR MILLING CORRESPONDENCE COURSE, *Milling*, 21 Sept. 1962.

GREER, E. N., English wheat for milling, *Milling* **139**: 480, 1962; English wheat for bread flour, *ibid.* **140**: 401, 1963.

GREER, E. N. and STEWART, B. A. *Classification of Home-grown Wheat Varieties*, 8th edition, Res. Assoc. Brit. Flour-Millers, St. Albans, Herts., 1965.

NATIONAL INSTITUTE OF AGRICULTURAL BOTANY, *Farmers' Leaflets*, N.I.A.B., Cambridge, annually.

PEACHEY, R. A., *Cereal Varieties in Great Britain*, Crosby Lockwood, London, 1951.

(*Opposite page*) FIG. 10. Part of a classification of English-grown wheat varieties according to their value for milling and baking purposes, by E. N. Greer and B. A. Stewart (The Research Association of British Flour-Millers, 1963). The characteristics in the four right-hand columns are graded on a scale of 1–4, lower figures indicating superior milling, bread- or biscuit-making qualities, and greater water absorption.

CHAPTER 5

WHEAT: THE FARM CROP

WHEAT QUALITY

Wheat passes through many hands between the field and the table: all who handle it are interested in the cereal, but in different ways.

The *grower* requires good cropping and high yields. He is not concerned with quality (provided the wheat is "fit for milling" or "fit for feeding") unless he sells grain under a grading system associated with price differentials (cf. p. 74).

The *miller* requires wheat of good milling quality—fit for storage, and capable of yielding the maximum amount of flour suitable for a particular purpose.

The *baker* requires flour suitable for making, for example, bread, biscuits or cakes. He wants his flour to yield the maximum quantity of goods which meet rigid specifications, and therefore requires raw materials of suitable and constant quality.

The *consumer* requires palatability and good appearance in the goods he purchases; they should have high nutritive value and be reasonably priced.

"Quality" in the general sense means "suitability for some particular purpose"; as applied to wheat, the criteria of quality are:

Yield of end product (wheat, for the grower; flour, for the miller; bread or baked goods, for the baker, etc.).

Ease of processing.

Nature of the end product: uniformity, palatability, appearance, chemical composition.

These criteria are dependent largely on environment—

climate, soil and manurial or fertilizer treatment. Within the limits of environment, quality is influenced by characteristics that can be varied by breeding, and is further modified during harvesting, farm drying, transport and storage.

GROWING

Since 1945 the acreage laid down to wheat in Britain has been stabilized by a system of subsidies, but the capacity for production has continued to increase by the use of new higher-yielding varieties, and by changes in husbandry. The ultimate aim of the grower is to obtain the maximum yield of "millable" wheat, just as it is of the plant breeder, even when he directs his attention towards varieties which are resistant to drought, frosts, diseases (Percival).

Both the yield and quality of the wheat crop are affected by conditions of soil, climate and farm management. The yield of flour obtainable from the wheat during milling is one criterion of wheat quality, and this is dependent on the degree of maturation—the extent to which individual grains are filled out with endosperm. Premature ripening, sometimes brought on by high temperature prevailing in the later part of the season, produces shrivelled grain, which is of high protein content because relatively more protein than starch is laid down in the endosperm during the early stages of ripening, whereas the reverse holds during the later stages.

Treatment with nitrogenous fertilizers is generally directed towards increase in yield, rather than increase in protein content; however, the effect of added nitrogen depends on the time of application and availability in the soil. Nitrogen taken up by the wheat plant early in growth results in increased grain yield, but taken up after heading it is laid down as additional protein in the seed, with consequent improvement of nutritive value and often of baking quality also. Possible ways of making nitrogen available at a late stage of growth are the early application of slow-acting fertilizers, or the late application of foliar sprays, e.g. urea, possibly by means of aircraft. However, belated nitrogenous fertilizing may encourage late tillering (branching by development of basal side shoots) and thereby increase the proportion of shrivelled grains.

DAMAGE TO WHEAT IN THE FIELD

Fungal diseases

Rusts. These are fungal diseases caused by species of the genus *Puccinia*. Yellow or Leaf Rust (*P. glumarum*) and Brown Rust (*P. triticina*) are occasionally troublesome in Britain. Black Rust (*P. graminis tritici*), sometimes occurring in the west of Britain, is particularly troublesome in the U.S.A., Canada and Argentina, and generally in countries with a hot climate.

Rusts exist in many physiological races or forms, and much research is carried out in America on the development of strains of cereals resistant to the various races. Immunity or susceptibility to rust is a varietal character, shown by Biffen (1907, 1912) to be inherited in Mendelian fashion. However, new races of fungal disease may arise to which hitherto resistant strains of cereals may be susceptible. This occurred with Thatcher wheat which was resistant to Stem Rust when released, but proved to be susceptible to race 15B in 1950 (cf. p. 71). Selkirk is a variety, bred for Canadian and U.S. HRS areas, which is resistant to Stem Rust race 15B.

Yellow or Leaf Rust is spread by air currents and attacks cereal plants in favourable weather in May and June in central and western Europe. Bright orange-yellow patches appear on the leaves; the patches increase in size and in number and eventually prevent photosynthesis occurring in the leaves, and the plant starves. In a bad attack, 80–90% of potential yield may be lost. Biffen bred wheat for immunity to leaf rust by using, as the immune parent, Rivet (*Triticum turgidum*), Club (*T. compactum*) or Hungarian Red (*T. vulgare*). The F_1 plants were all susceptible, but in the F_2 generation the ratio of immune to susceptible plants was 1:3. The character for immunity was thus recessive, but the F_2 immune plants bred true for immunity in the F_3 and subsequent generations.

Bunt, Stinking Smut. This is a disease caused by the fungus *Tilletia tritici*. The fungus enters the plant below ground, becomes systemic, and invades the ovaries. As the grain grows, it becomes swollen and full of black spores. Bunted grains are lighter than normal grains and can be separated by aspiration or flotation (see Ch. 6). Bunt imparts an unpleasant taint to the flour. The disease is satisfactorily controlled by mercurial seed dressings.

Loose or Common Smut. The fungus *Ustilago tritici* infects wheat plants at flowering time. The disease is of little importance to the miller, but is of concern to the grower because infected plants fail to produce seed. It can be controlled by treatment of seed wheat with hot water or hot formaldehyde.

Mildew. The fungus *Erysiphe graminis* infects the leaves of cereal plants during warm humid weather in April–June, later producing greyish-white patches of mildew. The leaf area becomes obliterated by patches of fungus, photosynthesis is reduced or prevented, and the plants become unable to develop normal grain. Even a mild attack reduces the yield. New varieties resistant to mildew in Britain are Swallow and Maris Badger (barleys), and Rothwell Perdix (wheat). However, most of the varieties of spring and winter wheat recommended by the National Institute of Agricultural Botany in Britain are moderately resistant to mildew.

Ophiobolus graminis and *Cercosporella herpotrichoides*, the fungi causing take all and eye spot diseases, live in the soil, and may survive on straw or stubble for a year or more. Plants affected by these diseases have empty or half-filled ears, and prematurely ripened or shrivelled grain.

Control. Chemical sprays are useful for control of some diseases, but they are not practicable or economic for control of rust, mildew, take all or eye spot. Rotation of crops should eliminate take all and eye spot, but mildew and rust can best be controlled by growing resistant varieties.

Animal diseases

Eelworm. Wheat may be attacked by the eelworm *Anguillulina tritici*, the grains becoming filled with the worms, which are of microscopic dimensions. Infected grains are known as "ear cockle" (not to be confused with "corncockle", the seeds of the weed *Angrostemma*).

Wheat Bug. Bugs of the genera *Aelia* and *Eurygaster* attack wheat plants and puncture the immature grains (see Fig. 11), introducing with their saliva a proteolytic enzyme which modifies the protein, preventing the formation of a strong gluten (cf. p. 177). Flour milled from buggy wheat gives dough that collapses and becomes runny if more than 5% of attacked grain is present. Steam treatment of the attacked wheat for a few seconds is beneficial in inactivating the enzymes, which are localized near the exterior of the grains. Wheat bug damage is generally restricted to crops grown in Russia, the Mediterranean littoral, eastern Europe and the Near East.

Wheat Blossom Midge. The damage caused by the midge *Sitodiplosis mosellana* varies greatly with year and locality. The female lays eggs in the wheat floret. The feeding larvae use part of the plant juices for their development; in consequence, infested grains become shrivelled. Secondary effects are reduced germination capacity and seed weight, and poorer baking quality of the flour.

Insect infestation of grain

It is estimated that about one-third of the world's production of wheat, maize and barley is infested by insects during the storage lifetime. The loss of wheat from insect ravages in the U.S.A. alone is valued at about 300 million dollars per annum. Damage due to insects is much less serious in Britain.

Certain insects live and complete the earlier stages of their life cycle entirely within the grains (cf. p. 130), which show no external signs of infestation. Such infested grains may, however,

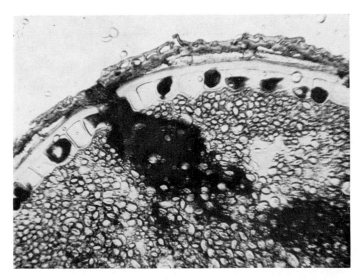

Fig. 11. Photomicrograph of section through wheat grain showing puncture by wheat bug. The dark mass is the injected secretion which fills the break in the pericarp and aleurone layer and extends into the endosperm. (Photo by J. J. C. Hinton.)

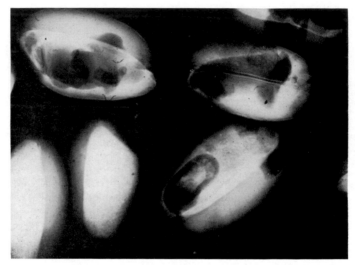

Fig. 12. X-ray photograph of six wheat grains, two uninfested (*bottom left*), four showing cavities caused by insect infestation. An insect is visible within the cavity of the grain at *bottom right*. (Part of a picture appearing in *J. Photogr. Sci.* **2**: 113, 1954; reproduced by courtesy of Prof. G. A. G. Mitchell and the Editor of *J. Photographic Science*.)

be detected by means of soft X-rays, which show the cavity in the grain and the insect within the cavity (see Fig. 12).

Harvesting

Sprouting in the ear. After the wheat seed appears to be "ripe", it needs a further period of maturation before it is capable of germination (cf. p. 200); during this period it is said to be dormant. The tendency to sprout, or germinate, in the ear depends both on varietal characteristics and on atmospheric conditions.

Hot dry weather hastens maturation; if followed by rain while the crop is still in the field, the conditions are favourable for sprouting. The wheat is less likely to sprout in a wet harvest if the season is cool.

Dormancy is a valuable characteristic conferring a degree of resistance to sprouting at harvest time. The factor appears to be related to the activity of certain enzymes, and to be linked genetically to redness of bran colour; white-branned varieties, such as Holdfast, tend to sprout in favourable conditions, whereas red varieties, such as Atle, do not.

Harvesting hazards. The condition of the harvested wheat stems from the harvest weather (Greer, 1963). Harvest hazards to wheat quality include:

(a) Cold weather, which may cause imperfect ripening, or delayed harvesting with blight infecting the chaff and spreading to the seed.

(b) Wet weather; if rain follows ripening, it may lead to premature germination.

Wheat harvested wet needs drying; mould may develop if drying is delayed or inadequate; overheating damage may ensue through rapid drying at too high a temperature, but may not be obvious until the flour, milled from the wheat, is baked.

Badly sprouted wheat is not of "millable quality", but mildly sprouted wheat, which may have an undesirably high enzymic activity (pp. 148 and 174), is often described as millable because the damage may not be visible.

Blighted wheat (wheat infected with moulds that discolour the flour) is still millable, although the flour quality will be inferior.

Combine harvesting. Since the advent of the combine harvester in 1943, when only 1500 machines were available in Britain, an increasing proportion of the wheat crop has been harvested by this means every year (cf. p. 131). The number of combines in use in Britain was 19,550 in 1952, and 57,000 in 1962.

When harvested by combine, the moisture content of the wheat should not exceed 15% for immediate storage, or 19% if the wheat can be dried promptly. Correct setting of the harvester to give efficient threshing coupled with the minimum of mechanical damage to the grain is important.

Discoloration by fungi. Mycelium of fungi such as *Aspergillus*, *Penicillium* and *Alternaria* is frequently present on and within the pericarp of sound wheat. In wet harvesting conditions, growth of the mycelium within the pericarp may be sufficiently prolific to cause discoloration and spoilage of the milled flour.

In the cold wet harvest of 1956 in England much of the wheat crop was invaded superficially by the fungus *Cladosporium*. Mycelium and sclerotia developed abundantly between epidermis and cross layer, and the flour colour was adversely affected through the presence of readily identifiable dark specks of fungal tissue. The infected grains contained no fungal tissue below the testa, but had dark vitreous patches in the endosperm because of imperfect development. Some improvement of flour colour resulted from repeated dry scouring of the wheat, followed by washing; colour of the flour was also improved by increasing the draught on the purifiers (cf. p. 123).

Storage

Wheat, even if the moisture content exceeds 19%, can be safely harvested by binder, stooked in the field, and stored in ricks, where it will dry with the minimum of deterioration. Wheat in ricks is frequently damaged by rats and mice: the average loss of grain from ricks in Britain, due to rodents, when no special precautions to exclude them have been taken has been estimated at 7%. Infestation of the ricks by rodents can be considerably reduced by baiting the ricks with Warfarin mixed with oatmeal (Dept. Agric. Scotland, 1958). The baits, placed in drain tiles, are inserted in the ricks during construction, and the residues can be removed during threshing (cf. p. 130).

Respiration. Combine harvested wheat and wheat threshed from ricks are stored in sacks or in bulk silos. The wheat grain is a living organism and as such is continuously respiring.

Respiration is slow at 14% m.c. and 20°C, but rises as moisture content and temperature increase. The respiration process itself generates heat, which is not quickly dissipated because wheat is a poor conductor of heat; thus, when wheat is stored at high moisture content, it respires rapidly, gradually warms up, and hence respires even more rapidly, the process being cumulative. In respiration the wheat gives off carbon dioxide and water vapour (thereby losing weight) and unless turned over, to allow evaporation of the water, will sweat and become caked in the bin.

Damp wheat (15·6–30% m.c.) is an ideal substrate for fungi; above 30% m.c. bacterial growth occurs, leading to spoilage, the production of much heat and possibly charring; the bacteria are killed when the temperature reaches about 60°C. Insect life also becomes more active as temperature rises and, because of their respiration, live insects present in grain also raise the grain temperature. If the moisture content of the wheat coming to the store (as grain) exceeds the critical safe moisture content, the wheat should be dried immediately.

Damp grain can be safely stored in air-tight bins. The inter-granular air is soon used up and respiration ceases. Sweating and heating also come to a stop. Grain so stored, however, is only fit for feed; thus this method of storage is not used in flour mills.

Deterioration during storage is promoted by mechanical damage during combine harvesting, because fungi attack damaged grains more readily than intact grains.

Moisture content for storage. The Ministry of Agriculture, Fisheries and Food has recommended maximum moisture contents for storage of sound grain (i.e. undamaged mechanically) at 65°F: 17% for 4 weeks' storage, 15% for storage until April, if in bulk or closely stacked bags without aeration. These moisture contents (which are the *highest* permitted in any part of a bulk, not *average* values) can be raised by 1% if the bulk store is aerated or the bags freely exposed to air. Corresponding maxima for seed grain or malting barley are 1% lower. Whatever its initial moisture content, however, the grain will slowly equilibrate with the atmospheric moisture, gaining moisture in conditions of high R.H., or losing it in a dry atmosphere. Regard must therefore be had to the prevailing atmospheric conditions in fixing safe moisture levels, remembering that ventilation in bulk bins is slower than in sacked grain.

All stored grain should be turned occasionally, but more frequently if the moisture content is near to the safe limit, or if the temperature of the grain begins to rise.

Grain drying. When drying damp grain, great care must be taken that the grain temperature be not allowed to rise too high. The moisture content of the grain, temperature and time of heating are all concerned in establishing safe drying conditions: with increasing moisture content, the maximum safe wheat temperature becomes steadily lower (for a given time of drying). The relationship between wheat moisture content and maximum drying temperature was investigated by Hutchinson (1944), whose results are shown diagrammatically in Fig. 14.

Excessive temperature causes damage to the protein, resulting in a lowering of baking quality (see Fig. 13). The germination capacity is reduced at temperatures somewhat lower than those that damage the gluten. Thus, when wheat at 14% m.c. was heated for 36 min, gluten was damaged by temperatures in the range 70–85°C, whereas germination was affected at temperatures of 64–72°C. Milling wheat should not, in any circumstances, be dried at temperatures above 65°C (150°F). A positive germination test may therefore be used as evidence that the baking quality

FIG. 13. Effect of heat damage during drying on breadmaking quality of wheat. A, loaf made from flour milled from untreated Atle wheat. B, loaf from same wheat after heating for 30 min at 161°F while at 15% m.c. C, loaf from same wheat heated at 168°F. (From E. N. Greer, *Milling* **127**: 318, 1956; reproduced by courtesy of the Northern Publ. Co. Ltd.)

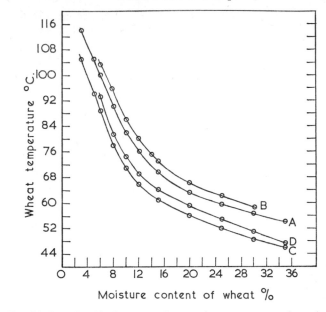

FIG. 14. Relationship between wheat moisture content and maximum safe drying temperature. Curves *A* and *B* correspond to nil germination after 60 min and 24 min heat treatment, respectively; curves *C* and *D* correspond to start of damage after 60 min and 24 min heat treatment, respectively. (From J. B. Hutchinson, *J. Soc. Chem. Ind.* **63**: 104, 1944; reproduced by courtesy of the Society of Chemical Industry.)

has not been damaged by heat treatment. Another useful test is made with the chemical tetrazolium (2,3,5-tri-phenyl-tetra-zolium chloride). The exposed surfaces of about 100 grains bisected longitudinally are soaked for 2 hr in the dark in a 0·2% aqueous solution of tetrazolium. The appearance of a red coloration at the germ end is proof of the existence of the enzyme dehydrogenase; non-appearance of the red colour indicates loss of viability.

WHEAT QUALITY—FOR THE MILLER

Having surveyed in the earlier part of this chapter the hazards that wheat may encounter during growing, harvesting and storage it is now possible to give the flour-miller's description of wheat of good quality from his point of view.

For the miller, good-quality wheat is:

1. *In good condition:* the grains are normal in colour (not discoloured) and bright in appearance; unweathered, free from fungal and bacterial disease, and unsprouted.

2. *Undamaged:* the grains are not mechanically damaged by the thresher, by insect infestation or by rodent attack, and have not been damaged by overheating during drying.

3. *Clean:* the grain is free from admixture with an abnormal quantity of chaff, straw, stones, soil, weed seeds and grains of other cereals, or of other types or varieties of wheat; the grain should be completely free from admixture with ergot, garlic or wild onion, bunt, rodent excreta and insects.

4. *Fit for storage*: the moisture content should not exceed 16% if for immediate milling, or 15% if the wheat is to go into storage (but see above, p. 90).

Besides these four aspects of quality—which are dependent mainly on the agricultural history of the wheat before the miller received it—the miller also wants the wheat to be of good milling quality, that is, to perform well on the mill: to give an adequate yield of flour, to process easily, and to yield a product of satisfactory quality.

MILLING QUALITY

The quality of wheat on the mill is measured by the yield and purity of the flour obtained from it. Good milling wheats, when properly conditioned (see Ch. 6) and milled under standard conditions, yield relatively more flour of lower ash content and lower grade colour (see p. 155) than poor milling wheats.

Purity of the flour means relative freedom from admixture with particles of bran. Bran is dark coloured whereas endosperm is white: the grade colour of the flour is an index of bran contamination. Another measure of bran contamination is the ash content, as there is much more ash in bran than in endosperm (cf. pp. 39 and 154).

The yield and purity of the flour are dependent on the way the endosperm separates from the bran when the wheat is ground; on the toughness of the bran, i.e. its resistance to fragmentation; and on the friability of the endosperm and the ease with which the flour is sifted. All these characteristics are related to grain texture and to the type of wheat (p. 67).

REFERENCES

BIFFEN, R. H. (1907, 1912), Studies in the inheritance of disease resistance, *J. Agric. Sci.* **2**: 109; **4**: 421.

DEPARTMENT OF AGRICULTURE FOR SCOTLAND (1958), *Rats and Mice on the Farm. Use of Warfarin*, D.O.A.S. Rodent Control, Edinburgh.

GREER, E. N. (1963), English wheat for bread flour, *Milling* **140**: 401.

HUTCHINSON, J. B. (1944), The drying of wheat, *J. Soc. Chem. Ind.* **63**: 104.

MINISTRY OF AGRICULTURE, FISHERIES AND FOOD (1963), *Preservation of Grain Quality during Drying and Storage*, H.M.S.O., London.

FURTHER READING

CORRESPONDENCE COURSE, *Milling*, 18 and 25 Jan. 1963.

GREER, E. N., Milling and baking tests on trial wheats, *J. Nat. Inst. Agric. Bot.* **7**: 291, 1955.

GREER, E. N. and HUTCHINSON, J. B., Dormancy in British-grown wheat, *Nature, Lond.* **155**: 381, 1945.

McCANCE, R. A. and WIDDOWSON, E. M., *Breads White and Brown*, Pitman Medical Publ. Co., London, 1956.

MORITZ, L. A., *Wheat and Flour in Classical Antiquity*, Clarendon Press, Oxford, 1958.

CHAPTER 6

WHEAT AT THE MILL: WHEAT CLEANING AND CONDITIONING

WHEAT CLEANING

Wheat arriving at the flour-mill may have been grown locally (home-grown) or may have been brought from another country (imported). The home-grown wheat will have come straight from the harvest field, or from storage in rick or silo, and it may already have been dried on the farm. The imported wheat will have passed from the field to a silo (or terminal elevator) where it may receive preliminary cleaning, and from there is brought by ocean vessel to a port and thence to the mill by barge, rail or road transport. Mills must therefore be equipped to take wheat in from all these types of vehicle, whilst mills stituated at ports are, in addition, equipped to take wheat in directly from ocean-going vessels (see Fig. 15).

The wheat, as the miller receives it, may contain impurities that enter from the field, during storage and transport, or accidentally. Those frequently encountered include:

Mud and dust

Weed seeds—both smaller and larger than the wheat grains

Other cereal grains—barley, oats, etc.

Straw and sticks

Husk (chaff)

Stones

Fungal impurities, e.g. bunt balls, ergot

Insects—both external to the grains and within the grain—and insect frass

Fig. 15. A large modern port flour-mill. (By kind permission of Spillers Ltd. Photo by courtesy of Henry Simon Ltd.)

FIG. 16. View in the screenroom of a large modern flour-mill showing two banks of Carter-Simon disc separators. (By kind permission of Hovis Ltd. Photo by courtesy of Henry Simon Ltd.)

Mites

Rodent excreta and hairs

String and binder twine

Fragments of metal—nails, nuts, etc.

These impurities, together with damaged, shrunken and broken kernels, are known collectively as Besatz on the continent of Europe.

The impurities must be removed before the wheat is milled. Some, like melilot seed and ergot, carry taints or are poisonous (cf. p. 232); others, like mud, discolour the flour and lower its quality. Stones and metal fragments present a fire hazard, and could damage the mill machinery. Other impurities, such as non-toxic weed seeds and other cereal grains, reduce the nutritive value of the flour or act as diluents. The processes of wheat cleaning are carried out in the part of the mill known as the screenroom in Britain, or cleaninghouse in the U.S.A. (see Fig. 17).

Impurities that adhere to the grain (mud, dust, hairs) are dealt with by washing, or by dry scouring, which loosens the impurities, coupled with aspiration, which lifts them away in an air current.

Impurities in the form of particles, unattached to the grains, are separated by machines which, in their operating principles, make use of characteristics in which the impurities differ from wheat. Such principles include differences in size (length and width), shape, terminal velocity in air currents, specific gravity, magnetic and electrostatic properties, colour, surface roughness, etc.

Wheat washing

In the wheat washing process, wheat is immersed in water (using 3–6 gal per 60 lb bu), conveyed by means of a worm to the base of a centrifugal machine called a whizzer, vigorously agitated, and spun-dried. During this process, there is a net gain of about 3% in moisture content.

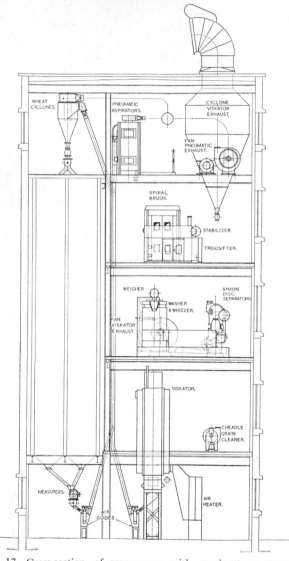

Fig. 17. Cross-section of screenroom with suction-type pneumatic elevators. (Reproduced from J. F. Lockwood, *Flour Milling*, 4th edition, Northern Publ. Co. Ltd. Liverpool, 1960, by courtesy of Henry Simon Ltd.)

Dry scouring

Wheat is fed into one end of a machine having either a perforated metal or an emery-lined cylinder, and is impelled against the inner side of the cylinder by rapidly revolving beaters, which also propel the grain towards the exit from the machine. Superficial dirt and loose shreds of beeswing (cf. p. 29) are removed by abrasion and blown away by air currents. This process removes hairs adherent to the grain.

Size and shape separations

Screens. Impurities larger or smaller than wheat grains are removed by screens of perforated metal, the size and shape of the perforations being chosen in relation to the grain size of the wheat. Round hole perforations of diameter approximately equal to the length of a wheat grain remove large impurities like maize grains, which overtail the screen, while the wheat passes through. Long-slotted perforations, narrower in small dimension than the width of the wheat grain, are used to remove small seeds and thin oats.

The screens are mounted in a frame which is caused to move horizontally, or nearly so, by gyrating or reciprocating motion. Wheat grains on a gyrating screen tend to remain horizontal and they separate by length, whereas on a reciprocating screen, provided the speed and amplitude of reciprocation are correct, the grains up-end and they separate by width. Screen sizes are therefore chosen appropriately to the method of screen movement. A third type of sieve motion is the Rexman motion: one end of a screen reciprocates while the other end gyrates. It is claimed that the apertures of the screen become less clogged with this type of motion.

Discs and trieur cylinders. Particles longer or shorter than wheat grains but similar in diameter can be separated from wheat by means of indented discs and trieur cylinders (see Fig. 16). The surfaces of the discs and the interior surfaces of the cylinders are provided with depressions or indents of carefully predetermined shape and size. Indents which are just deep enough to accommodate the impurities but too shallow for the wheat grains to enter as a whole are used to remove small impurities, such as round seeds, which are shorter than wheat grains. Indents just deep enough for wheat grains, but too shallow to accommodate large impurities, are used for separating wheat from impurities such as barley and oats, which are longer than wheat grains.

The action of discs and trieurs may be made more searching by re-feeding a cut-off from the tail of the machine back to the head for re-treatment, or by increasing the depth of the bed of stock: this improves the efficiency of separation. The capacity of discs is somewhat larger than that of trieur cylinders of equal diameter, but the selectivity of trieurs is said to be better.

In a typical arrangement of cylinders for separating large seeds, oats and barley from wheat, the mixed grain is first led to a splitter cylinder which lifts small wheat and large seeds, leaving large wheat, oats and barley. The two fractions are then led to other cylinders, one of which lifts the large seeds from the small wheat, whilst the other lifts large wheat from the oats and barley.

Helter-skelter or spiral seed separator. Particles differing in shape can be separated by allowing them to cascade down a tall spiral chute. Spherical particles attain a greater speed, and roll in a spiral of wider diameter than those which are ovoid (like wheat). The two types of particle can be collected in separate spouts by means of suitably disposed baffles.

Terminal velocity

Aspiration. The rate at which a particle falls in still air is the resultant of the speed imparted to it by the force of gravity acting on it, balanced by the resistance to free fall offered by the air. The resultant rate of fall, or "terminal velocity", depends on the weight of the particle and its surface area/volume ratio. Compact spherical or cubical particles thus have a higher terminal velocity than diffuse or flake-like particles. Instead of allowing the particles to fall in still air, it is more usual to employ an ascending air current into which the stock is introduced. The velocity of the air current can be regulated so that particles of high terminal velocity fall, while those of low terminal velocity are lifted.

Utilizing this principle, particles of chaff, straw, small seeds, etc., having a terminal velocity lower than that of wheat, can be separated from wheat in an aspirator, in which an air current is directed through a thin falling curtain of stock. In a *duo-aspirator* the stock is aspirated twice, permitting a more critical separation to be made. In this type of machine, the lifted particles are separated from the air current by a type of cyclone, and the cleaned air is re-cycled to the intake side of the machine.

The most efficient type of aspirator at present in use is the Twin-lift Aspirator (see p. 102).

Specific gravity

Differences in specific gravity between wheat and its impurities are utilized in the specific gravity separator, gravity table, or "air-float" machine. The machine consists of a triangular-shaped table or deck which is adjustably inclined to the horizontal both from back to front and from side to side, and capable of reciprocation from side to side. The rate and amplitude of reciprocation are also adjustable. A current of air from a blower is directed up through the table, which consists either of a fine mesh screen or of cintered porous metal. The air pressure over every part of the table should be equal, and of such intensity that the stock is kept in suspension, or "floated" over the table.

Stock is fed onto the table at the high point at the back of the machine and is floated towards the front of the deck, eventually spilling over into a series of hoppers. Individual particles, however, while travelling from back to front, move to left or right according to their specific gravity. Heavy particles tend to remain close to the surface of the table and move towards the higher side of the machine because of the thrust imparted to them by the reciprocation of the table. Lighter particles tend to float on the surface of the stock, and move towards the lower side of the machine because they receive less thrust from the reciprocation.

Although difficult to adjust, particularly when changing from one stock to another, the specific gravity separator in expert hands is capable of making excellent separations and can replace numerous conventional screenroom machines.

Colour

Separation by colour differences is possible by means of electronic colour-sorting machines. In one type of machine the particles of stock are projected into an optical box, and inspected one by one by photocells that view the particles against coloured backgrounds. Any particles that differ sufficiently in colour from the background are deflected from the main stream either by an air jet, or by receiving an electric charge and then being deflected by an electric field. The capacity of such machines is low when working on small-grained stock such as wheat, but the principle of colour separation could ideally be applied in conjunction with the operation of the loop system (q.v.).

Electrostatic separation

The separation of particulate material into fractions having differing electrical conductivity is accomplished at the electrodes of an electrostatic separator by means of high tension direct current electricity, commonly known as "static" electricity.

Material is fed into an electrostatic field created by a charged electrode discharging to an electrically grounded chute or roll. As the various particles enter the field, they assume different degrees of electrostatic charge, depending on their surface characteristics. Those taking the greater charge react most violently, i.e., are more attracted to or repelled from the electrode than the particles taking on a lesser charge. The two classes of particles those more and less heavily charged can then be isolated from each other by means of a mechanical splitter. It is rarely possible to obtain a precise separation from one pass through an electrostatic field of this type, hence electrostatic separators consist of numerous (sometimes 12) separating zones in series, and may include a built-in loop system enabling a rough cut to be made first, followed by recovery of fairly clean grain from the cut. The recovered grain can be re-fed to the head of the machine for re-treatment. The use of an electrostatic separator for removal of impurities (particularly rat and mouse excreta pellets) from maize has been described by Ake (1955).

Loop system

Although screenroom machines in theory deal with feed particles one by one, in practice the particles interfere with each other, and the efficiency of separation depends on machine design, feed rate, and proportion of cut-off (reject fraction). As the feed rate is reduced, interference between particles decreases, and efficiency of separation increases. As the proportion of cut-off increases, the rejection of separable impurities becomes more certain.

In the conventional screenroom arrangement, it is customary to feed to each machine, in succession, the entire feed, except for the small amount of screenings removed stage by stage. As the total removal of screenings frequently amounts to only $1-1\frac{1}{2}\%$ of the feed, every machine in the screenroom flow must have a capacity (i.e. be able to deal with a rate of feed) practically equal to that of the first machine. The feed rate can be reduced, and the efficiency of separation increased, by an arrangement known as the loop or by-pass system.

In the loop system (see Fig. 18), a large cut-off fraction (say 10%) is deliberately arranged so that it contains all the separable impurities, together with a proportion of clean wheat. The remaining 90% or so is accepted as clean. The cut-off is sent to a re-treatment machine for recovery of clean wheat. As the cut-off amounts to only about 10% of the total feed, the feed rate in the re-treatment machine can be much reduced, and the efficiency of separation improved.

Normal screenroom flow

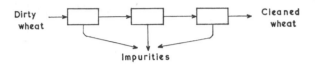

Loop or by-pass system

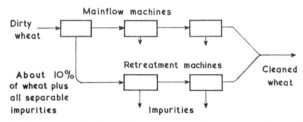

FIG. 18. Arrangement of machines in a normal screenroom flow (*above*), and in the loop or by-pass system (*below*). For explanation, see text.

The loop system is frequently applied in the section of the screenroom containing the discs and cylinders for length separation (see p. 97). The "cockle" (round seeds and small wheat) lifted out of the wheat is fed to a "re-cockle" machine, the purpose of which is to recover wheat from the "cockle". The same principle is also used in the Twin-lift Aspirator, which embodies main aspiration of the whole feed, and re-treatment of the cut-off to recover clean wheat.

CONDITIONING OF WHEAT

Objectives in conditioning

The objectives in wheat conditioning are primarily to improve the physical state of the grain for milling, and sometimes to improve the baking quality of the milled flour.

The processes of conditioning involve the addition of moisture to wheats that are too dry or its removal from those that are too wet, and often the heating and cooling of grain for definite periods of time, in order to obtain the desired average moisture content in the mass of the grain, and the desired distribution of moisture throughout each individual grain.

Milling objectives. The particular objectives of conditioning as regards milling are to toughen the bran and make it less brittle, to improve the separability of endosperm from bran, to increase the friability of the endosperm, and to improve the sifting of the flour. Other advantages follow when the wheat is correctly conditioned: when the bran is adequately toughened the flour is less contaminated with bran splinters and is thus whiter in colour and lower in ash content; when the endosperm is friable, the amount of power required to grind it is reduced.

Principles of conditioning

Effect of moisture. Both grinding and sieving, the principal operations in flour-milling, are affected by the moisture content of the stock. In general, as the moisture content of the wheat increases, the bran becomes tougher and less brittle, the endosperm becomes mellower and more friable, but cohesion between the bran and endosperm becomes stronger, so that the endosperm is less easily detached from the bran. As the moisture content of the ground stock increases, the separation of particles by sieving becomes more difficult.

Optimum moisture content. There is thus an "optimum" grain moisture content which will give the best milling results: high enough to mellow the endosperm and toughen the bran adequately, but not too high to hinder satisfactory cleaning of the bran and sieving of the stocks. The optimum moisture content varies for different wheat types, being higher for hard than for soft wheats. The moisture contents generally regarded as optimum are shown in Table 33.

It is customary in Britain to mill breadmaking flour from a mixed grist blended from hard, semi-hard and soft wheats; for ideal conditioning, each wheat of a mixed grist should be con-

TABLE 33

OPTIMUM MILLING MOISTURE CONTENTS

	%
Manitobas	16·5–17·5
Hard Red Winter	15·5–16·5
Plate	15·5–16·5
Soft Red Winter	15·5–16
English	15–16
Australian	15–15·5

ditioned separately to its own optimum moisture content. If a mixed grist were conditioned to a uniform moisture content, the bran of the soft wheat, being above its optimum moisture content, would not clean up adequately, whereas that of the hard wheat, at too low a moisture content, would tend to shatter. In practice, the hard wheat group is generally conditioned separately from the soft group, and the two blended immediately before the first break of the milling process.

However, in establishing the optimum moisture content for milling, there are other factors to be considered:

(a) *Extraction rate:* for milling high extraction "white" flour (80–85%), the optimum moisture content is $1–1\frac{1}{2}\%$ lower than the optimum for milling flour of 70% extraction rate (cf. p. 126).

(b) *The required moisture content in the products:* the moisture content of the products is generally lower than that of the mill feed, on account of evaporation of moisture from the intermediate products during milling. The extent of evaporation is dependent on the moisture content of the stock and the atmospheric conditions within the milling plant, and usually amounts to $1–2\frac{1}{2}\%$. The miller will generally aim at a moisture content of about 14% in the finished flour, because flour of this moisture content can be safely stored for some months. However, if a lower moisture content were specified, e.g. in certain service contracts, the moisture content of the millfeed would have to be reduced accordingly.

Moisture movement in the grain. In order to bring wheats to their optimum moisture content, dry wheats are moistened by the addition of water (or steam): damp wheats are dried. Either process causes a moisture gradient to be set up in the grain.

When wheat is moistened, the outer layers of the bran quickly absorb moisture, but the testa (seed coat) is relatively impervious and hinders the diffusion of moisture from bran to endosperm except in the vicinity of the germ, where the testa layer is interrupted. Thus, when wheat is immersed in water, moisture rapidly enters at the germ end and more slowly over the remainder of the surface.

Recently damped wheat is moist on the outside but still relatively dry inside; the converse holds for recently dried wheat. If damped or dried wheat is allowed to rest at normal temperature (e.g. 60–65°F) for some days, the moisture gradient gradually decreases. This is illustrated by figures in Table 34.

TABLE 34

MOISTURE CONTENTS OF BRAN AND ENDOSPERM OF DAMPED AND
DRIED WHEATS

Rest period	Damped from 13 to 14·5%		Dried from 17·7 to 14·5%	
	Bran m.c. (%)	Endosperm m.c. (%)	Bran m.c. (%)	Endosperm m.c· (%)
1 hour	15·6	14·2	13·8	14·9
1 day	14·7	14·4	14·0	14·8
1 week	14·5	14·5	14·1	14·8
1 month	14·5	14·6	14·1	14·8

Effect of heat. The rate of movement of moisture in the grain is slow at normal temperature but increases as the temperature rises; this principle is utilized both in the drying of wheat and in warm conditioning (which is the inverse of drying, and is intended to drive the external moisture into the interior of the grains).

Heat is used in conditioning for two purposes: (1) to accelerate movement of moisture in the grain which would take place more slowly in the absence of applied heat; (2) to produce a beneficial effect on the baking quality of the flour from wheats of certain classes. The first is achieved at wheat temperatures up to 115°F, and the process of treatment is called "warm conditioning";

the second requires wheat temperatures higher than 115°F and is called "hot conditioning".

Wheat is conveniently heated by contact with hot water radiators or by hot air. Water is better than air as a conductor of heat: its specific heat is 1·0 as against 0·238 for air. Furthermore, 1 lb of water occupies 0·016 ft³ as against 14 ft³ for 1 lb of air; hence, much smaller volumes of water than air are required to carry the same amount of heat.

Drying. Wet wheat may be dried by heating it and then by drawing cold air through it in *open* circuit. The air, which at low temperature has a low water-holding capacity, is heated by the grain, its water-holding capacity is correspondingly increased and it leaves the grain laden with moisture. Alternatively, warm dry air, at a carefully controlled temperature, may be drawn through the heated grain. If hot air is blown through cold wheat in order to dry it, there is some danger of overdrying the surface and forming a hard skin which impedes diffusion of moisture from the interior and may even cause deterioration of baking quality. However, if the warmed wheat, dried as described with cold air, is finally cooled rapidly by drawing large volumes of cold air through it, the surface of the grain tends to be dried, while the interior remains moist.

Conditioning. If, after raising its temperature, the wheat is held at the desired temperature in a closed container, or if air with a suitably high relative humidity is blown through it in *closed* circuit, the average moisture content will remain constant, but the moisture distribution within the grains will tend to alter. If the wheat has not been recently damped (i.e. bran drier than the endosperm), moisture will be drawn out from the interior of each grain and will condense on the surface (a phenomenon known as "sweating") until a certain moisture distribution, in which the bran is damper than the endosperm, has been established. If the wheat has been recently damped, so that the bran is much damper than the endosperm, much of the external moisture will be driven into the interior in these circumstances, until the bran is only slightly damper than the endosperm, and the final moisture distribution between bran and endosperm will be similar to that in the former case. During cooling, the surface of the grain will lose moisture, so that the moisture distribution in the cooled wheat will be fairly uniform.

As the ideal moisture distribution in the millfeed is achieved when the bran is slightly damper than the endosperm, a second light damping is sometimes given within 1–2 hr of milling, so that the added moisture remains in the bran.

Baking quality of the flour. This is measured by the water-absorbing capacity (cf. p. 175), and the power to produce large, well-shaped loaves (Simon). The latter depends on gas production, gas retention, and the physical characteristics of the gluten (extensibility and resistance) (cf. p. 149 and Ch. 10).

There is little evidence that warm conditioning (wheat temperature $\not> 115°F$) produces any change in starch, protein or other flour constituents different from that obtained by conditioning in the cold. It has been claimed, however, that at wheat temperatures of 150–160°F, reached in hot conditioning, some improvement in baking quality results, particularly in wheats of low stability. Swanson *et al.* (1916) found improvement in the loaf vol. from flour milled from Kansas wheat that had been damped to 13·5% m.c. and heated to 190°F for 3–12 min. Severe heat treatment of wheat tends to make the gluten harsher and tougher, i.e. less elastic and extensible. This is a desirable effect in soft wheats, but not in hard wheats; hence, Manitoba wheat is seldom heated beyond 110°F. The baking quality can easily be ruined by heat treatment that is too severe, and hence it has been recommended that heat treatment, if required, should be applied to the flour rather than to the wheat (see p. 139).

Time. Moisture equilibration throughout the grain takes a long time at normal temperature, and resting times of 24–72 hr after damping have been advocated. Resting periods of this length require much bin space, which is both expensive and inconvenient. The main advantage of warm conditioning is that rate of moisture entry into the grain, and moisture distribution within the grain, can be so accelerated that the process taking

24 hr in the cold can be accomplished in $1\frac{1}{2}$ hr in the warm conditioner.

Theories about the mechanics of conditioning

The processes of conditioning presumably involve mechanical, physico-chemical and biochemical principles.

Enzymic activity is stimulated by increased moisture and by elevated temperature, but normal conditioning moistures, temperatures and times do not affect diastatic activity. Proteolytic activity is not affected by temperatures used in warm conditioning, although it may be reduced by hot conditioning treatment.

Expansion and contraction. The grain swells when moistened, and it may also expand slightly when the temperature is raised; corresponding contractions occur on drying and possibly on cooling. Milner and Shellenberger (1953) showed by means of X-rays that cracks appear in the endosperm under certain conditions of damping and drying, and suggested that the cracks relieve strains set up in the endosperm due to expansion and contraction (cf. p. 66).

According to this theory, the ideal milling condition can be regarded as one in which a reticulum of cracks has been induced, such that the dimensions of the tessera are approximately those of flour particles, and the wheat, upon milling, falls into pieces along the lines of the cracks.

Form of moisture. Water is said to exist in the grain in two forms: partly as free moisture and partly as moisture of constitution. The latter may be chemically bound, probably to the protein. It is believed that the degree of hydration of the protein may influence baking quality.

Methods of conditioning

Blending of wheats. If damp and dry wheats are blended and allowed to lie together, moisture transfer from the damper to the drier grains occurs. However, equilibration does not reach completion within a measurable time. Furthermore, the moisture gradient in the originally damper grains in the blend is such that the bran coats are drier than the endosperm (cf. Table 34, p. 105), which is undesirable in the millfeed. This primitive method of conditioning is therefore seldom used in the modern mill.

Cold conditioning. The moisture content of wheat that is too dry may be increased by adding the required amount of cold water and keeping the wheat moving until the moisture has been absorbed. For an addition of up to 3%, absorption is complete in a few minutes. The wheat washing process itself adds $3-3\frac{1}{2}\%$ of moisture. If an addition of more than about 3% is required, it is preferable to damp the wheat stepwise, with frequent turning over and aspiration between successive dampings. After moistening, the wheat must lie for one to three days at normal temperature to allow the moisture, held superficially, to diffuse into the interior of the grains.

It has been claimed that the penetration of moisture into the grain in cold conditioning can be hastened by small concentrations of additives, such as Aerosol O.T. (used by Sullivan, 1941) and sodium bicarbonate (used by Altrogge, 1955), and that these have no adverse effect on the grain.

Warm conditioning. To avoid the delay of 1–3 days' resting time in cold contioning (for moisture equilibration), the damped wheat may be "warm conditioned" for $1-1\frac{1}{2}$ hr at wheat temperatures up to 115°F. It is recommended, however, that warm conditioned wheat be rested for 24 hr before milling.

The many types of modern conditioners are fully described in larger textbooks (e.g. Lockwood, Smith). Figure 19 shows in diagrammatic form the construction of a typical conditioner.

For warm conditioning, the washed wheat from the whizzer (wet on the outside) goes directly to the heating section of the conditioner (see diagram) where its temperature is raised quickly to the desired optimum by contact with hot water radiators. The wheat falls to the conditioning section, where it remains about 1 hr while some of the external moisture is forced

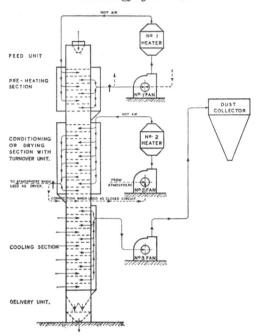

Fig. 19. Vertical section through Universal conditioner dryer. (Reproduced from J. F. Lockwood, *Flour Milling*, 4th edition, Northern Publ. Co. Ltd., Liverpool, 1960, by courtesy of Henry Simon Ltd.)

into the interior either by heat from radiators (the wheat being in an enclosed space) or by hot, moist air circulating through the wheat in closed circuit. Finally, the wheat passes down to the cooling section, where it is cooled by cold air currents in open circuit.

Hot conditioning. The procedure for hot conditioning is modified from that described for warm conditioning, so that the wheat temperature rises to 140°F or higher, but is maintained thereat for a shorter period of time. Hot conditioning is less frequently practised than warm conditioning because of the risk of ruining the baking quality of the gluten by over-treatment.

Steam conditioning. The transfer of heat to the wheat grains is much more rapid from steam than from hot air or from radiators, so that even shorter times of treatment become possible by the direct application of steam to heat the grain and moisten it at the same time. Correspondingly, the time and temperature ranges for safe treatment become narrower.

In comparison with normal conditioning, steam conditioning requires less power; steam conditioned wheat is said to give a higher yield of bran and a higher yield of flour of quality equal to or better than that obtained by other methods of conditioning.

Altrogge (1957) has described a differential steam-addition method for conditioning various types of wheat. His recommended temperatures in the steaming, heating and cooling sections, for three types of wheat, are shown in Table 35.

TABLE 35
ALTROGGE'S RECOMMENDATIONS FOR STEAM CONDITIONING

	Wheat types		
	Hard, dry	Soft, dry	Damp
Temp. at end of steaming section (°F)	100/108	131/149	104/113
Temp. in heating section (°F)	104	104	140
Temp. in cooling section (°F)	104→77	104→77	140→77

Besides the usual milling improvements claimed for warm conditioning, Altrogge claimed that in his steam method the effect of the highest temperature was advantageously restricted to the outer layers of the grain.

Vacuum drying as a step in conditioning. Drying under vacuum results in the rapid removal of moisture at safe temperature. For example, the B.P. of water is 33°C (91°F) at 0·05 atm pressure (95% vacuum). A conditioning method that has been recommended by Dienst (1954) consists in steaming grain at 1·5 atm pressure to increase the moisture content to 2–3% higher than that required for milling, and then drying the wheat back at 0·05 atm pressure and 91°F. The damped grain is said to become mellowed without shrinkage during the vacuum drying. The use of high pressure steam reduces the time required for the initial swelling of the grain and is said to loosen the bran. This process, however, although it is said to result in higher yields of flour, a saving in power consumption, and to be safe in operation, does not appear to have been developed commercially to any extent, except perhaps in Germany.

Conditioning for air-classification

Besides the objectives mentioned above (p. 102), a further objective in the conditioning of wheat might be improved response to fine grinding and air-classification (cf. pp. 142–146). Recent reports by Stringfellow *et al.* (1963, 1964) showed that protein displacement was increased when the moisture content at which HRW wheat was milled was raised from 14 to 16%, or when the wheat was pre-treated by steaming at 150°F for 15 min, or by subjecting to alternate cycles of freezing (at 0°F) and thawing while at 40% m.c., or by freezing in liquid nitrogen while at 30% m.c. Kent (1965) found that protein displacement was increased when *flour* moisture content was reduced before fine grinding (cf. p. 142).

REFERENCES

AKE, J. E. (1955), Electrostatic grain cleaning, *Bull. Assoc. Oper. Millers* 2188.

ALTROGGE, L. (1955), Versuche mit dem Netzolit-Verfahren, *Die Mühle* **92**: 590.

ALTROGGE, L. (1957), Neues zur Dampfkonditionierung, *Die Mühle* **94**: 558.

DIENST, K. (1954), Sechs Varianten der vollständigen und kontinuierlichen Weizen-Konditionierung im Vakuumverfahren einschliesslich der Randzonenbehandlung, *Müllerei* **7**: 113.

KENT, N. L. (1965), Effect of moisture content of wheat and flour on endosperm breakdown and protein displacement, *Cereal Chem.* **42**: 125.

LOCKWOOD, J. F. (1960), *Flour Milling*, 4th edition, Northern Publ. Co. Ltd., Liverpool.

MILNER, M. and SHELLENBERGER, J. A. (1953), Physical properties of weathered wheat in relation to internal fissuring detected radiographically, *Cereal Chem.* **30**: 202.

SMITH, L. (1944), *Flour Milling Technology*, 3rd edition, Northern Publ. Co. Ltd., Liverpool.

STRINGFELLOW, A. C., PFEIFER, V. L. and GRIFFIN, E. L., JR. (1963–4), Air classification response of flours from Hard Red Winter Wheats after various premilling treatments, *Northwestern Miller* **269** (13): 12; **270** (1): 12.

SULLIVAN, B. (1941), Quick tempering of wheat for experimental milling, *Cereal Chem.* **18**: 695.

SWANSON, C. O., FITZ, L. A. and DUNTON, L. (1916), The milling and baking quality and chemical composition of wheat and flour as influenced by: 1, different methods of handling and storage; 2, heat and moisture; 3, germination, *Kansas Agr. Exp. Sta. Tech. Bull.* 1.

FURTHER READING

BRADBURY, D., HUBBARD, J. E., MACMASTERS, M. M. and SENTI, F. R., *Conditioning Wheat for Milling*, Misc. Publ. No. 824, U.S.D.A., Washington, D.C., 1960.
CORRESPONDENCE COURSE, *Milling*, 1 March 1963.
SCHÄFER, W. and ALTROGGE, L., *Wissenschaft und Praxis der Getreidekonditionierung*, Schäfer, Detmold, Germany, 1960.

CHAPTER 7

WHEAT AT THE MILL: FLOUR-MILLING

OBJECTIVES

The objectives in the milling of white flour are:

1. To make, as completely as possible, a separation of the endosperm from the bran and germ, so that the flour shall be free from bran specks and of good colour, and so that the palatability and digestibility of the product shall be improved and its storage life lengthened.

2. To reduce the maximum amount of endosperm to flour fineness, thereby obtaining the maximum extraction of white flour from the wheat, and at the same time to ensure that the amount of damage to the starch granules does not exceed the optimum (cf. p. 175).

The reduced endosperm is the flour; the germ, bran and residual endosperm is a by-product (wheatfeed in Britain; millfeed in U.S.A.) used primarily in animal feeding.

DEVELOPMENT OF FLOUR-MILLING

In prehistoric times, the barley and husked wheat (emmer: *Triticum dicoccum*) used for human food were dehusked by pounding the grain in mortars. The invention of rotary grain mills, for grinding ordinary bread wheats (*T. vulgare*), is attributed to the Romans in the second century B.C. Thereafter, until the development of the rollermill in the mid-nineteenth century, wheat was ground by stone-milling. In western Europe the local soft wheat was ground by a "low grinding" process, in which the upper stone was lowered as far as possible towards the lower stone, thereby producing a heavy grind from which a single type of flour was made.

Where contrasting social conditions existed side by side, as in eighteenth-century France and nineteenth-century Austria, the use of flours of more than one quality became possible. For this purpose, a "high grinding" system was used, with the upper stone raised, producing a gritty intermediate material from which, by further treatment, flours of diverse quality could be made.

Hard wheat, from the Danube basin, was ideally suited to the high-grinding system, and, when steam power became available, Hungary became the centre of the milling industry. It was there that the new rollermilling system was developed: in 1860 the first complete rollermill was operated in Pesth, and, thereafter, the rollermilling/high-grinding system was taken to other countries.

PROCESSES OF FLOUR-MILLING

In order to separate the endosperm from the bran and germ and reduce the endosperm to flour fineness, a particular form of grinding has been adopted which is a combination of shearing, scraping and crushing, achieved by rollermills, and which exploits the differences in properties between the endosperm, bran and germ. It is essential to minimize the production of fine particles of bran, and this basic requirement is responsible for the complex arrangement of the modern flour-milling system and for the particular design of the specialized machinery used in it, and also for the conditioning process described in Ch. 6.

There are three basic processes:

Grinding: fragmenting the grain or its parts, with some dissociation of the anatomical parts of the grain from one another.

Sieving: classifying mixtures of particles of differing particle size into fractions of narrower particle size range. Particular sieving processes include: *scalping:* sieving to separate the break stock (the coarsest particles) from the remainder of a break grind; *dusting, bolting, dressing:* sieving flour from coarser particles; *grading:* classifying mixtures of semolina, middlings and dunst into fractions of restricted particle size range.

Purifying: separating mixtures of bran and endosperm particles, according to terminal velocity, by means of air currents.

Gradual reduction process

The modern rollermilling process for making flour is described as a "gradual reduction process" because the grain and its parts are broken down in a succession of relatively gentle grinding stages rather than by one extremely severe grinding stage, as in the now superseded process of stonemilling. No attempt is made to achieve either of the stated objectives (p. 114) completely in a single grinding stage; the severity of grinding is carefully adjusted so that only the required amount of endosperm fragmentation and bran cleaning occurs at each stage.

STOCKS AND MATERIALS

The various intermediate stocks and products in flour-milling are described as follows:

Feed: material fed to, or entering, a machine.

Grind: the whole of the ground material delivered by a rollermill.

Break stock: the portion of a break grind overtailing the scalper cover (sieve), and forming the feed to a subsequent grinding stage. The corresponding fraction from the last break grind is the *bran.*

Break chop or *break release:* the throughs of the scalper cover, consisting of semolina, middlings, dunst and flour.

Tails, overtails: particles that pass over a sieve.

Throughs: particles that pass through a sieve.

Aspirations: light particles lifted by air currents.

Tins: aspirations lifted in purifiers and deposited in the "tins".

Semolina (Sizings): endosperm in the form of coarse particles (pure, or contaminated with bran or germ), derived from the break system.

Middlings (*Break Middlings*) : endosperm, intermediate between semolina and flour in particle size and purity, derived from the break system.

Dunst: this term is used to describe two different stocks: (1) Break dunst: endosperm finer than middlings, but coarser than flour, derived from the break system. This stock is too fine for purification but needs further grinding to reduce it to flour fineness. (2) Reduction dunst (reduction middlings, in the U.S.A.): endosperm, similar to middlings in particle size, but with less admixture of bran and germ, derived from semolina by rollermilling.

Flour: endosperm in the form of particles small enough to pass through a flour sieve, e.g. one having 100 meshes per linear inch.

FLOUR-MILLING OPERATIONS IN BRIEF

A succession of grinding stages, using rollermills in each (see Fig. 20), is used to open the whole grain in the first stages, and, in subsequent stages, to grind parts of the grain selected from the products of preceding grindings.

An individual grinding stage may be light or heavy, according to the character of its feed: in general, feeds consisting of relatively pure endosperm can be heavily ground, whereas those containing much bran must be lightly ground.

Each grinding stage yields a "grind" consisting of a mixture of particles of varying sizes, and is followed by a sieving process in which the particles are sorted into two or more fractions according to particle size.

Particles differing widely in size (in the grind from any one stage of rollermilling) also differ in composition: particles of endosperm, which tend to be friable, are generally smaller than particles of bran, which tend to be tough and leathery. The process of sieving, besides sorting the particles according to size, thus to some extent separates particles of differing composition from each other, e.g. endosperm from bran.

Each grinding stage makes a proportion of flour (the smallest particles of the grind), which is removed in the subsequent sieving operation to form part of the end product. Each grinding stage also yields a proportion of coarse particles of one or both of two kinds: (a) potential flour-yielding particles: these are conveyed to a subsequent grinding stage; (b) particles which can yield no useful flour: these are removed from the milling system and contribute to the by-products (offals) as "bran" or "wheat-feed" (millfeed, in the U.S.A.).

The succession of grinding stages is grouped into three systems: break, scratch and reduction. Each has a distinct objective.

The *break system* consists generally of four or five "breaks" or rollermill grinding stages, each followed by a sieving stage. The system is fed with the whole grain in the first break, and with the break stock in subsequent stages. The objective in the first break is to open the whole grain, and in subsequent breaks to scrape off the endosperm from the bran coats which, owing to their fibrous structure and tough nature, remain in relatively large pieces.

Break rollermills

Each of the break rollermills is equipped with a pair of rolls, usually 9 or 10 in. in diameter and 40 in. long, mounted diagonally (in Britain) or horizontally (in the U.S.A.) in parallel alignment along their whole length. The gap, or "nip", between the aligned rolls can be adjusted to allow precision grinding. The rolls rotate in opposite directions so that the surfaces of both rolls are entering the nip in the same direction. One of the rolls rotates at a faster speed than the other, the speed differential between the two generally being $2\frac{1}{2}$:1 for break rollermills (see Fig. 22).

The break rolls are corrugated or grooved over their entire surfaces, the corrugations (or "flutes") running along the length of the roll, but disposed at a slight angle to their axes, this angle being known as the "spiral". In Britain a spiral of 1 in 7 (i.e. the corrugations move 1 unit around the roll for each 7 units

FIG. 20. Rollermill floor of a large modern flour-mill. (By kind permission of Moinho Paulista Ltd. Photo by courtesy of Henry Simon Ltd.)

FIG. 21. Plansifter floor of a large modern flour-mill. (By kind permission of Joseph Rank Ltd. Photo by courtesy of Henry Simon Ltd.)

Break rolls

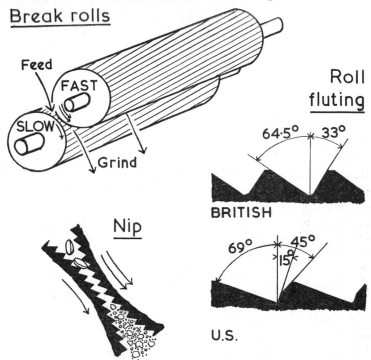

Fɪɢ. 22. Pair of break rolls showing flutes, with enlarged view of the "nip". Details of typical British and U.S. break roll flutings are shown *right*.

they move along the axis) is common, whereas in Canada and the U.S.A. a shallower spiral (e.g. 1 in 24) may be used.

In section, each flute resembles an italic V, with one side shorter and steeper than the other. The flutes of the two rolls forming a working pair are so arranged that at the nip the flutes cross each other at an angle double that of the spiral of a single roll. The fast roll is said to be in the "sharp" disposition if the steeper side of the V enters the nip before the shallower side, the opposite disposition being termed "dull". For the slow, or holding, roll, the steep side of the flute entering the nip first gives the dull disposition. The disposition of the pair of rolls, generally stated by giving that of the fast roll first, e.g. dull to sharp, has a considerable influence on the grinding effect. New rolls are frequently run dull to dull, and subsequently changed to dull to sharp or sharp to dull, when the initial sharpness of the flutes has worn off.

The effect of the speed differential between the two rolls and the spiral arrangement of the flutes is to cause the flutes of the fast roll to move across those of the slow roll at the point of contact, simulating the action of scissors.

The stock to be ground is fed into the nip in the form of a thin curtain. In the first break, the flutes shear open the grain, often along the crease, and unroll the bran coats so that each consists of an irregular, relatively thick layer of endosperm closely adpressed to a thin sheet of bran (see Fig. 5, and cf. p. 23). A small amount of endosperm is detached from the bran coats, mostly in the form of chunks of up to about 1 mm^3 in size ("semolina"); small fragments of bran are also broken off, but little flour is made.

The first break grind thus consists of a mixture of particles which differ in size and composition. The largest are the bran coats (break stock), still thickly coated with endosperm; the intermediate-sized particles are either semolina, middlings and dunst (large- and medium-sized particles of pure endosperm) or bran snips (some are free bran, some are bran loaded with endosperm); the smallest are flour, i.e. tiny particles of endosperm. The particles are separated from one another by scalping, dusting and grading.

The bran coats (break stock), the overtails of the wire scalping sieve, are fed to the second break. The grinding process is similar to that of the first break except that the flutes on the surface of the rolls are slightly finer, and thus more closely spaced, and the working distance between the two rolls at the nip (the "roll gap") is somewhat less. Typical flutings and roll gaps for the various breaks, quoted by Scott (1951), are shown in Table 36.

The feed to the subsequent breaks consists of the break stock scalped from the grind of the previous break; in each grinding

TABLE 36
TYPICAL BREAK ROLL FLUTINGS AND GAPS

Break	Roll fluting (flutes per in.)	Roll gap (in.)
1st	12	0·020
2nd	14	0·006
3rd	18	0·0035
4th	26–30	0·003

stage endosperm is scraped from the bran coats, which become progressively thinner and better cleaned. The roll gap is made progressively narrower and the flutes progressively finer at each successive break. The coarse fraction scalped from the last break grind can yield no more useful endosperm by further roller-milling, and forms the by-product "bran".

Break release

Typical break releases (throughs of the scalper cover), as percentages of the individual break feeds, and of the feed to 1st break, and the usual scalping covers (sieves) employed in white flour milling, as quoted by Lockwood (1960), are shown in Table 37.

TABLE 37
TYPICAL BREAK RELEASES AND SCALPING COVERS

Break	Break release			Scalping cover	
	Percentage of feed to break	Percentage of 1st break feed	Total release	Wire number	Aperture length (mm)
1st	30	30	30	20 w.	1·00
2nd	52	36	66	20 w.	1·00
3rd	35	12	78	28 w.	0·71
4th	14	3	81	36 w.	0·53

It is generally considered desirable to release on the breaks about 10% more stock than is required in the straight-run flour, subsequently rejecting the 10% from the scratch and reduction systems as fine offal.

The break release is dressed over a flour silk to separate the flour (break flour) which is removed from the milling system as part of the end product. The coarser part of the break release (semolina, middlings, dunst) is graded by sieves into a number of fractions differing in particle size, but all consisting mainly of endosperm with some admixture of bran and germ. The semolinas and middlings are then purified (cf. p. 123) to remove particles of beeswing (cf. p. 29) and free bran, and to separate particles of loaded bran from the bulk of the material which is pure endosperm.

The fractions consisting of loaded bran are fed to the scratch system (cf. p. 123), the purified semolina and middlings to the reduction system (cf. p. 124).

Sieving

Screening or bolting cloth is made of woven wire, silk or nylon; wire is generally used for scalping, silk or nylon for dressing, dusting and grading, although wire may also be used for these purposes.

The sieving process is carried out in either plansifters or centrifugals (see Fig. 21). The plansifter is a machine consisting of a vertical nest of horizontal sieves, the whole assembly gyrating in a horizontal plane. The mixture to be sieved is fed in at the top and falls from sieve to sieve by gravity. A single plansifter may incorporate sieves of four or five different mesh sizes, thereby delivering five or six fractions of differing particle size. A certain depth of stock on the sieving cloth is required to obtain a satisfactory sieving effect. Owing to the movement of the sieve, the stock resting on it becomes stratified; the finest particles work down until they are adjacent to the cloth, whilst the largest ride up to the surface of the stock. The overlying layers of medium and coarse particles thus tend to keep the finest particles in contact with the sieve cloth and assist their passage through the sieve apertures.

Centrifugals are long polygonal reels, covered entirely (except at the ends) with bolting cloth, and rotating on a horizontal axis which passes centrally along the length of the reel. The stock to be sifted is fed into the reel at one end and is flung centrifugally against the cloth by beaters or impellers. The particles passing through the cloth are collected inside the casing that surrounds the moving parts; the overtails gradually work along to the opposite end of the machine where they are discharged.

Centrifugals are being replaced by plansifters in modern mills. For equivalent sieving capacity, plansifters occupy less space than centrifugals.

Purifying

A purifier consists of an oscillating sieve set at a slight angle to the horizontal, and enclosed by a hood or cover. The sieve comprises four sections which become progressively coarser in size of mesh from head to tail. The hood, divided into sections to correspond to the sieve sections, is connected by trunking to the suction side of a fan so that air, controlled by valves, is drawn up through each sieve section. Stock to be purified is fed onto the head end of the sieve and is moved down by the oscillating motion towards the tail or coarse sieve section. Particles of endosperm, decreasing in purity from head to tail sieve sections, fall through the respective sieves, against the direction of the air current (being relatively dense and having high terminal velocity), and are collected in hoppers; light particles of free bran and beeswing are lifted by the air current and are collected in trays ("tins") suspended above the sieves. Particles intermediate in density and terminal velocity, consisting of bran loaded with endosperm, are not lifted, nor do they fall through the sieves, but remain floating on the sieve, and are eventually discharged as overtails.

Scratch system

The scratch system, in British mills, consists of two–four grinding stages (denoted by X, Y, Z). The feed contains large semolina and particles of bran with attached endosperm, but the average particle size is much smaller than that of break stock. The purpose of the scratch system is to scrape bran from the endosperm without undue production of fine stock, and to reduce oversize semolina. Fluted rolls are used, but the rolls are fluted more finely than break rolls, e.g. 50 cuts per inch. The grind from the scratch rolls is scalped, dressed, graded and purified in much the same way as that from the breaks, and the fractions despatched respectively to subsequent scratch grinding stages, to offals, to "flour" (as end product), or to the reduction system, according to their character and particle size.

In pre-war milling the scratch system was important, and it was developed even more during the war-time milling of long-extraction white flour in Britain. With the present-day trend towards shorter extractions and simplified mill diagrams, the scratch system tends to be curtailed, and in some mills may no longer be found. There is no separate scratch system in U.S. mills.

Reduction system

This consists of 12–15 grinding stages interspersed with siftings for removal of the (reduction) flour made by each preceding grind, and of a coarse (tailings) fraction from some of the grinds. The nomenclature of the reductions in Britain and in the U.S.A. is as follows:

Britain	U.S.A.	Britain	U.S.A.
A	Sizings	F	1st tailings
B	1st middlings	H	5th middlings
C	2nd middlings	J	2nd tailings
B_2	2nd quality	K	6th middlings
D	3rd middlings	L	7th middlings
E, G	4th middlings		

Reduction systems in U.S. mills generally have some additional stages beyond the 7th middlings.

Grinding is carried out on rollermills, but these differ from the break rollermills in two important respects:

(a) The roll surfaces are quite smooth, or slightly roughened (matt surface); fluted rolls are rarely used and then on one or two particular reductions only.

(b) The speed differential between the two rolls is lower, usually $1\frac{1}{4}$:1 in Britain ($1\frac{1}{2}$:1 in the U.S.A. and Canada).

These characteristics of the reduction rollermill result in the grinding effect being a crushing rather than a shearing action. The feed to the reduction system consists of the purified semolina and break middlings, and the dunst, together with suitable fractions from the scratch system. The coarse and medium semolinas feed A and B rolls, respectively, the objective in these grindings being to reduce the semolina to dunst. The latter, after grading and dressing, together with purified break middlings of corresponding particle size, goes to C roll. The coarser fractions sieved from A roll grind are sometimes purified, but this is not general practice; products of the other reduction rolls are not normally purified.

The feeds to A and B rolls, despite careful purifying, inevitably contain a few particles of bran; the feeds also contain particles of germ. By carefully controlling the roll pressure, these bran and germ particles can be flattened without fragmentation, and subsequently sifted off as coarse "cuts" from the grinds. These, and corresponding fractions from subsequent reduction stages, are fed to certain reduction stages, collectively known as the "coarse reduction system" or the "tailings system". In Britain these stages are generally designated B_2, F and J rolls. The other reduction stages, viz. D, E, G, H, K and L, regrind the dunsts from the preceding grinding stages, although small quantities of new material of low grade, derived from the break or scratch systems, may be introduced at suitable points from D roll onwards in order not to contaminate the feeds to A, B and C rolls, which yield the purest or patent flour. In the longest reduction systems, M is an additional coarse roll, and N, O and P are additional fine rolls.

Figure 23 shows a simple mill flow in diagrammatic form; the flow of a larger mill would be similar in general arrangement, but would show more grinding stages, with more detailed sub-division of the intermediate stocks.

LONG EXTRACTION MILLING (cf. pp. 160–163)

The process described above, for milling white flour of, say, 68–72% extraction, is modified for milling "white" flour of higher, e.g. 80–85% extraction in various ways, as follows:

(a) Releases are increases throughout the break system, by narrowing the roll gap, and by using scalping covers of slightly more open mesh.

(b) Operation of the purifiers is altered by adjusting the air valves, altering the sieve clothing and re-flowing the cut-offs so that less stock is rejected to fine wheatfeed, and more goes to the scratch and reduction systems.

(c) The scratch system is extended by the use of additional grinding stages.

(d) The extraction of flour from the reduction system is increased by more selective grinding, and by employing additional grinding stages to regrind offaly stock that would be rejected as unremunerative in white flour milling.

(e) Additional flour may be obtained by changing the flour silks to slightly more open sizes.

(f) Still additional extraction can be obtained, at the sacrifice of good flour colour, by bringing the wheat onto the 1st break rolls at $1\frac{1}{2}$–1% lower moisture content than the optimum for white flour milling.

RECENT DEVELOPMENTS IN FLOUR-MILLING

In an attempt to push up the extraction of white flour and obtain improved separation of endosperm from bran and germ, milling systems tended to become more complex, with sub-division of stocks and multiplication of grinding stages, until early in this century.

Since the last war there has been a marked trend towards simplification, characterized by a decrease in the number of grinding

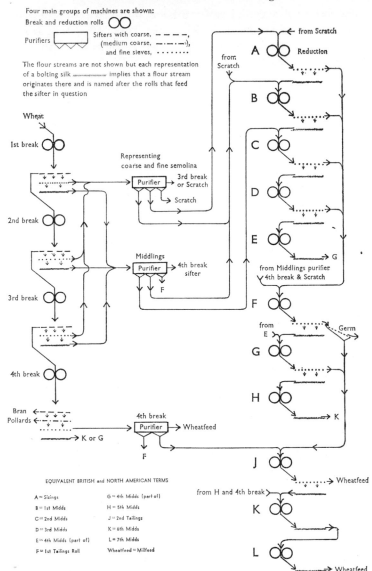

FIG. 23. Diagrammatic representation of flour-milling flow. (Reproduced from C. R. Jones, *Proc. Nutr. Soc.* **17**: 9, 1958, by courtesy of the Cambridge University Press.)

stages, faster roll speeds, heavier loading on rollermills and plansifters, and elimination of purifiers (see Table 38). As a result, the grinding capacity of mills has been considerably increased without increasing the size of the mill.

TABLE 38

ROLL SPEEDS AND SPECIFIC ROLL LENGTHS, USED NOW AND
PREVIOUSLY IN VARIOUS COUNTRIES *

Country	Roll speeds†		Specific roll length‡ (in./sk/hr)
	Break (rev/min)	Reduction (rev/min)	
Britain, pre-war	350	210	110–120
Britain, 1946	350	210	80–85
Britain, 1963	450	400	60–65
North America, 1908	475		42
North America, 1963	550	500	30–33
Western Europe, 1963			35
Hungary, 1963			21
U.S.S.R., C.S.S.R., 1963			16

* Source of data: Jones (1964).

† Speed given is that of the fast roll.

‡ Total length of roll contact per unit weight of straight-run flour produced per hour.

The *Knauff* system (Germany) uses special roll flutings, speed differentials of 6:1 or 7:1 on I and II Bks., few grinding stages, and heavy loading.

The *Meyer* system uses special roll flutings and various roll speeds.

The *Donath* system employs means for varying roll speed differential during operation.

The *Pratique* system (France, U.S.S.R., C.S.S.R.) uses very heavy loading of grinding rolls, and low specific roll length (20 in./sk/hr).

The *Bellera* or *Molinostar* process (U.S.A., U.S.S.R., C.S.S.R.) and the *Bühler* system (Germany) use low specific roll lengths (35 in./sk/hr), heavy loading of plansifters, but no purifiers.

The *MIAG intensive* system dispenses with purifiers, restricts the number of reduction grinding stages, but employs bran dusters and disintegrators depending on vibration and impact; the latter feature also characterizes systems developed recently by British milling engineers.

WHOLEMEAL AND BROWN FLOURS (cf. p. 151)

These are flours or meals having extraction rates higher than 85%: they are frequently made by adding all (for wholemeal) or some (for brown flour) of the offals to the straight-run flour, milled as white flour. The coarse bran would generally be ground before blending with the flour and fine offals. Alternatively, the whole grain may be ground between stones in one or two stages, to form a coarse wholemeal or, if a small amount of the coarsest bran is removed, a coarse brown flour. Certain proprietary brown flours are made by hammer-mill or impact grinding of the wheat without any separation or sieving stage.

GERM

The germ is not an intentional constituent of white flour, although a proportion of germ inevitably finds its way into the flour. Whole germs (embryos), of the size of coarse semolina, are split off from the bran coats in the break process, and go via the purifiers to the feeds of A and B reduction rolls. On these reductions the germs are flattened, with little fragmentation, and are subsequently sifted off as A and B roll tails, going thence to B_2 roll and subsequently to F and J rolls. At any of these stages it is possible to make a separation from the grind, by means of suitably sized sieves, of material which consists largely of germ. The germ may be collected as a separate by-product, or may be flowed with the remainder of the fine offals to wheatfeed.

FLOUR-MILL PESTS

The common animal pests of flour-mills and warehouses are mice and rats, insects and mites.

Mice and rats can be controlled by careful attention to the structure of the buildings to render them rodent-proof, and then by dealing with any infestation by conventional methods of trapping and baiting. Warfarin mixed with oatmeal (0·005% for the brown rat *Rattus norvegicus*, 0·025% for the mouse *Mus musculus*) can be used in mills if adequate safety precautions are taken (cf. p. 89). Newly erected buildings can easily be rodent-proofed, but old and dilapidated buildings are more difficult and expensive to proof.

Warfarin is an anticoagulant, based on hydroxycoumarin. An improved rodent bait contains, beside the anticoagulant, the substance sulphaquinoxiline, a bactericide. The latter immobilizes bacteria in the digestive system which secrete vitamin K, an antidote to the anticoagulant, and improves the effectiveness of the Warfarin.

The principal insect pests of flour-mills are beetles and moths.

Beetles

Sitophilus granarius (formerly called *Calandra granaria*) (grain weevil) and *S. oryzae* and *S. zeamais* (rice weevils). The adults are about 4 mm long, shiny black or brown, with elongated snout. The female lays eggs inside the grains. The larvae live on the endosperm and pupate inside the grain. The adult eats its way out of the grain. *S. granarius* adults are inactive at 50°F but can lay eggs at 55°F; *S. oryzae* is more sensitive to the cold.

Rhizopertha dominica (lesser grain borer). The adults are about 1 mm long, with a relatively large head, shiny dark brownish black in colour. Active multiplication occurs above 70°F, and the insect is a serious pest in hot countries.

Laemophleus pusillus (formerly called *L. minutus*; flat grain beetle) and *Cryptolestes ferrugineus* (formerly *L. ferrugineus*; rust red grain beetle). The adults are 1·5–2·5 mm long and much flattened. They attack sound wheat, but are also found in milled products and in mill machinery.

Tribolium confusum (confused flour beetle) and *T. castaneum* (rust red flour beetle). The adults are about 3·5 mm long, red-brown in colour, with no snout. The larvae are about 4 mm long, dirty white in colour. They are found in damaged wheat and in mill machinery.

Oryzaephilus surinamensis (saw tooth grain beetle). This has recently increased in importance as a pest of home-grown grain in Britain. The widespread use of combine harvesters (cf. p. 88) involves more artificial drying of grain, and the warm humid atmosphere of the grain drier provides an ideal environment for the pest, which does not breed at temperatures below 65°F. It attacks broken grain and milled products, but not whole grain. The adult is flat, brown, and about 3 mm long.

Two larger beetles are found in mills: (1) *Tenebroides mauritanicus* (cadelle), 8–13 mm long and wasp-waisted; the larvae are 17 mm long, greyish-white with black head. The cadelle lives on other insects, but also attacks grain and its milled products, and damages silk sieving cloths. (2) *Tenebrio molitor* (meal worm), 13–17 mm long, dark brown or black. The larvae are 25–31 mm long × 5 mm broad, light brown in colour. They eat grain products and other insects, but do not attack whole grain. The meal worm is found in dark, damp crevices, between floor boards, in disused spouts.

Moths

Anagasta (formerly *Ephestia*) *kühniella* (mill moth or Mediterranean flour moth). This is the best-known flour-mill pest. The moth is grey in colour, about 12 mm long. The larvae, which are white in colour, feed on ground stocks, but also on whole wheat if at a fairly high moisture content. They spin webbing which binds together particles of stock, causing chokes.

Ephestia elutella (cacao moth). This is one of the most destructive pests of stored grain. The eggs are laid loose on the surface of the grain. The larvae penetrate the germ, in which they live during two stages of development. Subsequent stages are passed outside the grain, the larvae continuing to feed on the germ.

When fully grown, the larvae migrate to the surface of the grain mass, leaving a trail of webbing, and thence to the ceiling of the bin or store, where they rest until the following May before pupating.

Sitotroga cerealella (angumois grain moth). The adult moth is about 12 mm long, buff in colour, the hind wings grey. The larvae, which are about 5 mm long, white with yellow heads, bore into grain and feed on the endosperm.

Mites

Acarus siro (formerly *Tyroglyphus farinae*; grain or flour mite). This mite attacks wheat and milled products. It bores into the grain and feeds within, imparting a characteristic smell described as "minty". Although resistant to low temperature, mites are sensitive to moisture content and cannot develop in cereal products with less than 13% m.c.

HYGIENE

Insects do not flourish if frequently disturbed and exposed to light; hence, regular sweeping and cleaning help to control them. In cleaning the mill, particular attention must be given to dead spaces in machinery, elevators, etc., and to the segregation of returned bags that may be infested.

Rate of insect multiplication increases up to 70°F and can be reduced by regular turning over and aspiration of cereals, thereby lowering the temperature.

Disinfestation

The need for disinfestation is particularly acute in warm countries, where insects breed rapidly.

Heat treatment

Heat treatment of grain, e.g. at 145°F for 2 min, 135°F for 10 min, or 125°F for 30 min, is effective in destroying insect life, but care must be taken that the gluten is not damaged by excessive heat treatment (cf. p. 90).

Entoletion

Wheat containing live insects can be sterilized by passage through an entoleter (cf. p. 141) run at about 1750 rev/min. Hollow grains and insects may be broken up and can be removed by subsequent aspiration.

Air-tight storage of grain

In air-tight storage of grain (cf. p. 89), e.g. in underground pits, the carbon dioxide given off by insect respiration accumulates and eventually kills all insect eggs and larvae.

Fumigation

Insect life is destroyed by fumigation, which is carried out in mills at regular intervals. Ethylene dichloride and carbon tetrachloride are used for fumigating grain; ethylene oxide for flour; methyl bromide for flour-mills, grain and milled products— although methyl bromide is not used on flour in Britain because it imparts a slight taint. Adequate safety precautions should be taken with all fumigants.

Insecticides

Insecticides used as dusts or smoke generators include pyrethrum, for all food premises including bulk flour bins; gamma BHC or DDT smoke generators for silos containing insect-infested grain, and for empty contaminated silos, warehouses and bags. Gamma BHC is recommended for dealing with mites; dieldrin may be used against spider beetles and cockroaches, but only when the necessary precautions to avoid contamination of foodstuffs can be observed.

Ionizing radiation

This is another promising tool for insect control and prevention of infestation. Gamma-radiation, e.g. from Co^{60}, penetrates the grain and effectively kills insect eggs. Wheat in India infested with rice weevil showed no fresh emergence of insects within 2 months of gamma-radiation treatment. Feeding studies on rats, mice, dogs and poultry showed no hazards from consumption of cereals irradiated at dosage levels required for disinfestation. The nutritive value of such treated cereals was not impaired. The U.S.A. has recently approved irradiation treatment of wheat for export (*American Miller*, 1963).

In order to achieve uniform dosage levels it is desirable to treat grain while it is moving through the disinfestor in a stream; moreover, the electron accelerators used for producing gamma-radiation require adequate screening by a considerable thickness of concrete, and it is generally more convenient to bring the grain to the source of radiation than vice versa. The cost of transport then greatly exceeds the cost of the disinfestation process itself. The installation of gamma-radiation disinfestors would thus appear to be feasible only at ports continuously handling large tonnages of grain.

REFERENCES

ANNOTATION (1963), Irradiation effective—and expensive, *Amer. Miller* **91** (9): 5.

JONES, C. R. (1964), Recent European developments in flour milling technology, *Milling* **142**: 324.

LOCKWOOD, J. F. (1960), *Flour Milling*, 4th edition, Northern Publ. Co. Ltd., Liverpool.

SCOTT, J. H. (1951), *Flour Milling Processes*, 2nd edition, Chapman & Hall, London.

FURTHER READING

JONES, C. R., The essentials of the flour-milling process, *Proc. Nutr. Soc.* **17**: 7, 1958.

KENT, N. L., SIMPSON, A. G., JONES, C. R. and MORAN, T., *High Vitamin Flour*, Ministry of Food, H.M.S.O., 1944.

SIMON, E. D., *The Physical Science of Flour-Milling*, Northern Publ. Co. Ltd., Liverpool, 1930.

SMITH, L., *Flour Milling Technology*, 3rd edition, Northern Publ. Co. Ltd., Liverpool, 1944.

CHAPTER 8

FLOUR

FLOUR GRADES

Flour is produced by every grinding machine in the break, scratch and reduction systems of the normal mill flow. The stock fed to each grinding stage is distinctive in composition—in terms of proportions of endosperm, bran and germ contained in it, and the region of the wheat grain from which the endosperm is derived—and the quality of each "machine" flour is correspondingly distinctive. The numerous machine flours differ widely in baking quality, colour and granularity, and in contents of ash, fibre and nutrients.

In Britain there are today no recognized standards for flour grades: each miller makes his grades according to customers' requirements, and exercises his skill in maintaining regularity of quality for any particular grade.

If the flour streams from all the machines are blended together in their rational proportions, the resulting flour is known as "straight run grade". Other grades are produced by selecting and blending particular flour streams in particular proportions. Frequently the ash content or colour grade of the flours is used as a guide for grouping them in order to obtain grades of predetermined average ash content or average colour grade.

Table 39 shows typical yields and ash contents of flour streams from a well-equipped and well-adjusted mill making flours of fairly low ash content.

BLEACHING

Bleaching of the natural pigment of wheat endosperm by

TABLE 39
TYPICAL YIELDS AND ASH CONTENTS OF FLOUR STREAMS*

Flour stream	Typical yield (%)	Ash content (%)	Group
C, A	21·6	0·30 ⎫	⎫ 1
B, D	14·8	0·35 ⎬	⎬
B₂, E, F	7·5	0·38–0·40	1 or 2
G, H	5·7	0·42–0·45 ⎫	⎫
I, II Bk.	3·55	0·47 ⎬	⎬ 2
I, II Bk. Cs. Midds.	5·0	0·49 ⎭	⎭
I, II Bk. Med. and Fine Midds.	4·0	0·51–0·52 ⎫	⎫
J, K	3·1	0·53 ⎬	⎬ 3
III Bk. CMD	1·05	0·54 ⎭	⎭
L	1·6	0·62 ⎫	⎫
M, N	1·5	0·70–0·75 ⎬	⎬ 4
IV Bk. Cs., O	1·9	0·90 ⎬	⎬
IV Bk. Fine, P	1·1	1·0 ⎭	⎭
Total and average	72·4	0·42	

* Data condensed from Lockwood (1960).

oxidation occurs rapidly when flour is exposed to the atmosphere, more slowly when flour is stored in bulk. The bleaching process can be greatly accelerated by treatment with chemicals; those permitted in Britain are chlorine dioxide and benzoyl peroxide, the former also acting as an "improver".

The endosperm pigment was formerly thought to be carotene, the precursor of vitamin A, and flour colour was expressed in "carotenoid" units. The pigment is now known to be about 95% xanthophyll or its esters, of no nutritional significance. The carotenoid unit for expressing flour colour was nevertheless retained until colour grading by electrical meters, such as the Kent-Jones & Martin Colour Grader, became customary (cf. p. 155).

Chlorine. Chlorine gas is ideal for treatment of cake flour and its use with this type of flour is permitted in Britain. The usual dosage is 3–6 oz/sk. The Bread and Flour Regulations 1963 do not permit its use with bread flour in Britain.

Nitrogen trichloride. This gas, known as "Agene", replaced chlorine in the early twenties as an improving and bleaching agent for breadmaking flour because it was much more effective. Its use has been discontinued from 1955, after it had been shown by Mellanby (1946) that flour treated with Agene in large doses might cause canine hysteria (although Agene-treated flour has never been shown to be harmful to human health).

Bentley *et al.* (1950, 1951) found that nitrogen trichloride reacts with methionine, an amino acid in wheat protein, to form a toxic derivative, methionine sulphoximine.

$$\underset{\text{Methionine}}{\overset{\displaystyle H}{\underset{\displaystyle NH_2 \qquad H}{COOH\text{-}CH\text{-}CH_2\text{-}C\text{-}S\text{-}CH_3}}} \qquad \underset{\text{Methionine sulphoximine}}{\overset{\displaystyle H \; O}{\underset{\displaystyle NH_2 \qquad H\,NH}{COOH\text{-}CH\text{-}CH_2\text{-}C\text{-}S\text{-}CH_3}}}$$

Chlorine dioxide. Chlorine dioxide (Dyox) has now replaced Agene in Britain, the U.S.A. and Canada as the most widely used improving and bleaching agent. The gas is produced by passing chlorine gas through an aqueous solution of sodium chlorite. The chlorine dioxide gas is released by passing air through the solution, and is applied to breadmaking flour at a rate of 2 g/sk. The Bread and Flour Regulations 1963 (effective from September 1964) state that the chlorine dioxide gas used in Britain may not contain more than 20% of chlorine by volume. Chlorine dioxide treatment of flour destroys the tocopherols (cf. p. 58).

Benzoyl peroxide. This is a solid bleaching agent (Nvaodelox) supplied as a mixture with starch. The dosage rate, normally 10–20 ppm, is restricted to 50 ppm in Britain by the Bread and Flour Regulations 1963. The bleaching action occurs within about 48 hr. This bleacher has the advantage over gaseous agents that only a simple feeder is required, and storage of chemicals presents no hazard; the fact that it has no improving action is advantageous in the bleaching of patent flours. The treated flour contains traces of benzoic acid, but objection has not been raised.

Acetone peroxide. This is a dry powder bleaching and improving agent, marketed in the U.S.A. as Keetox, a blend of acetone peroxides with a starch carrier. Its use is permitted in the U.S.A. but not, as yet, in Britain. It may be used alone, or in combination with benzoyl peroxide. The usual dosage rate is 2 oz of Keetox per 280 lb sk of flour.

MATURING OR IMPROVING

An improvement in baking quality occurs slowly during storage of flour. This change, known as maturation or "ageing", can (like bleaching) be accelerated by chemical substances or "improvers", which modify the physical properties of gluten during fermentation in a way that results in better bread being obtained. Matured flour differs from new flour in having better handling properties, increased tolerance in the dough to varied conditions of fermentation, and in producing loaves of larger volume and more finely textured crumb.

Chlorine and chlorine dioxide both bleach and improve, whereas benzoyl peroxide acts as bleacher only. Improving agents, permitted in Britain from September 1964, that do not bleach are potassium bromate, potassium persulphate, ammonium persulphate, acid calcium phosphate and ascorbic acid (vitamin C). Such substances nevertheless give a whitened appearance to the loaf because of their beneficial effect on the texture of the crumb. Improving agents do not increase the carbon dioxide production in a fermented dough, but they improve gas retention (because the dough is made more elastic) and this results in increased loaf volume (cf. p. 176).

Action of improvers

The action of improvers is believed to be an oxidation of the cysteine sulphydryl or thiol (—SH) groups present in wheat gluten. Consequently, these thiol groups are no longer available for participation in exchange reactions with disulphide (—S:S—) groups—a reaction which is considered to release the stresses in dough—and consequently the dough is tightened, i.e. the extensibility is reduced. Alternatively, it has been suggested that the oxidation of —SH groups may lead to the formation of new —S:S— bonds which would have the effect of increasing rigidity. High-speed mixing of dough in oxygen also results in improvement of gluten characteristics; this effect has been ascribed by Hawthorn and Todd (1955) to the action of unsaturated fat oxidases, and the direct uptake of oxygen by the protein. For further discussion see Pace (1959) and Hlynka (1964).

Ascorbic acid is used as a flour improver in the U.K., Germany and elsewhere. It can be oxidized to dehydroascorbic acid (DHA) through catalytic action of ascorbic acid oxidase, or by atmospheric oxygen. The DHA is the component responsible for the improving action. An enzyme "DHA reductase" is required for oxidation of sulphydryl (—SH) compounds by DHA.

Azodicarbonamide (ADA) is a new flour maturing agent, marketed as Maturox. When mixed into doughs it oxidizes the sulphydryl (—SH) groups and exerts an improving action. Oxidation is rapid and almost complete in doughs mixed for $2\frac{1}{2}$ min. The residue left in the flour is biurea. ADA could possibly replace iodate in dough processes where faster maturation is required. Flour treated with ADA is said to produce drier, more cohesive dough than that treated with chlorine dioxide, to show superiority in mixing properties, and to tolerate higher water absorption. An average treatment is 10 ppm of azodicarbonamide. The agent does not bleach, but the bread from treated flour appears whiter because of finer cell structure. The use of azodicarbonamide is permitted in the U.S.A. but not as yet in Britain.

The breadmaking quality of flour can be improved by physical means, e.g. by controlled heat treatment (see pp. 107, 110), but such treatment is not regarded as a practical substitute for chemical improvers.

Another physical improving process is the aeration process, in which flour is whipped with water at high speed for a few minutes and the batter then mixed with dry flour. Improvement is brought about by oxidation with oxygen in the air, probably assisted by the lipoxidase enzyme (cf. p. 49) present in the flour. A similar improving effect can be obtained by overmixing normal dough (without the batter stage): cf. the Chorleywood Process (p. 182).

FLOUR BLENDING FOR BLEACHING AND IMPROVER TREATMENT

Because the various flour streams differ in quality, the optimum level of bleaching and treatment varies correspondingly, the lower grade flours (those nearer the tail end of the break and reduction systems) in general requiring more treatment. As it is not practicable to treat each machine flour separately, it is customary to group the machine flours into three or four streams, according to quality. A possible grouping is indicated in Table 39 (p. 136). Each group would be given appropriate bleacher and improver treatment: e.g. the lowest 20% might receive treatment at ten times the rate for the best quality 80%. The final grades are then made up by blending two or more of the groups in desirable proportions.

STORAGE OF FLOUR

Flour is stored in bags made of jute, cotton or paper, or in bulk bins. Bags of flour in Britain usually contain 140 lb when packed; the bags are stacked, often several tiers high, on palletting.

The hazards to flour in storage include those to wheat in storage, viz. mould and bacterial attack, and insect infestation (cf. p. 130), and also oxidative rancidity (cf. p. 49) and eventual deterioration of baking quality. The optimum moisture content for the storage of white flour is 13%. At moisture contents higher than 13%, mustiness, due to mould growth, may develop, even if the flour does not become visibly mouldy. At moisture contents lower than 13%, the risk of fat oxidation and development of rancidity increases. The reactions involved in oxidative rancidity are catalysed by heavy metals ions, such as Cu^{++}.

Freedom from insect infestation during storage can be ensured only if the flour is free from insect life when put into store, and if the store itself is free from infestation. Good housekeeping in the mill and the milling of clean grain should ensure that the milled flour contains no live insects, larvae or eggs, but as a precautionary measure flour is often passed through an entoleter before being bagged off. The entoleter is a machine consisting of a rotor rapidly rotating within a fixed housing; the flour is fed in centrally and is flung with considerable impact against the casing. At normal speeds of operation (2900 rev/min for flour) the machine is extremely effective for the destruction of all forms of insect life and of mites, including eggs (cf. p. 132). The insect fragments, however, are not removed from the flour by the entoleter.

Heavy jute or twill bags for packing flour are returnable and must be cleaned before re-use. Despite the most stringent precautions, returned bags present an infestation hazard unless they are sterilized. Non-returnable bags made of paper—plain, or laminated and impregnated—are more expensive and less convenient to handle, but their use eliminates, entirely, any infestation hazard.

Bulk storage and delivery of flour

The storage of flour in bulk bins, and delivery in bulk containers, has advantages over conventional storage and delivery in bags. Although constructional costs of bulk storage facilities are high, the running costs are low because manhandling is much reduced, and warehouse space is better utilized. About one-fifth of all the flour milled in Britain is delivered in bulk.

Bins for storing flour in bulk are made to hold 70–100 tons each. Packing pressure inside the bin increases with bin area, not bin height; a bin area of 60 ft^2 is satisfactory. Concrete bins are better than wooden bins, which are liable to become infested; the inner surfaces must be smooth, and the corners rounded. Bins may be filled and emptied pneumatically; compressed air for fluidizing the flour has proved successful for discharging. Bulk wagons for transport can be filled by gravity feed at the mill, and discharged by a blowing system direct to the storage bins at the bakery.

PROTEIN DISPLACEMENT MILLING

The contents of the cells of wheat endosperm are principally starch and protein; the starch is in the form of granules (cf. p. 31) which are embedded in a protein matrix. The interstitial protein, in the form of approximately triangular pieces, between the starch granules, has been described by Hess (1955), from its shape, as Zwickelprotein (wedge protein). He described as Haftprotein (held protein) the protein which, he claimed, occurred as a sheath around the starch granule (see Fig. 24).

Flour as normally milled consists of a mixture of particles differing in size and composition (cf. p. 67). Jones *et al.* (1959) classified these into three main fractions: (1) whole endosperm cells (singly or in clumps), segments of endosperm cells, clusters of starch granules and protein, and large detached starch granules (upwards of 40 μ in diameter); (2) medium-sized starch granules, some with protein attached (15–40 μ in diameter); (3) small chips of protein, and detached small starch granules (less than 15 μ in diameter). The protein content of fraction 1 is usually similar to, or slightly higher than, that of the parent flour; the protein contents of fractions 2 and 3 are respectively $\frac{1}{2}$–$\frac{2}{3}$ and twice that of the parent flour. Thus, particle size separation at appropriate "cut sizes" gives some degree of protein separation, making it possible to obtain from one parent flour fractions of varying protein content, thereby increasing the range of usage of the flour (see Fig. 25).

Fine grinding. The proportion of medium-sized and small particles (below 40 μ) in flour, as normally milled, is about 45% by weight in soft wheat flour, but only about 20% in hard; it can be increased (and the proportion of the large particles reduced) by a process of fine grinding (see Table 40). One type of machine

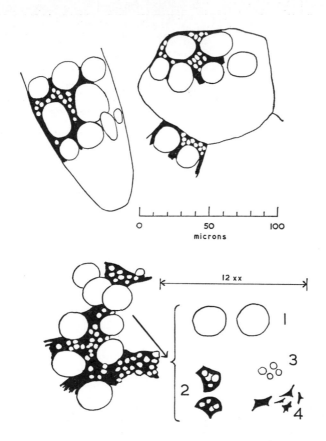

FIG. 24. *Above:* the two main types of endosperm cell—prismatic (*left*), polyhedral (*right*)—showing large and small starch granules (white) embedded in protein matrix (black). *Below:* exposed endosperm cell contents (*left*) and products of further breakdown: 1, detached large starch granules (about 25 μ diameter); 2, "clusters" of small starch granules and protein matrix (about 20 μ diameter); 3, detached small starch granules (about 7 μ diameter); 4, fragments of free wedge protein (less than 20 μ diameter). 12xx is the representation to scale of the mesh aperture of a typical flour bolting silk. (Redrawn from C. R. Jones *et al., J. Biochem. Microbiol. Technol. Engng.* **1:** 77, 1959, and reproduced by courtesy of Interscience Publishers.)

1	2	3	4	5	6
			Types of particle		
Size range μ	Yield % by wt.	% protein	Detached starch granules	Free wedge protein	"Clusters" also, above 28μ, cell segments
0–13	4	19			
13–17	8	14			
17–22	18	7			
22–28	18	5			
28–35	9	7			
Over 35	43	11·5			
Initial protein: 9·5%					
Weighted average protein for total under 35μ: 8·1%					

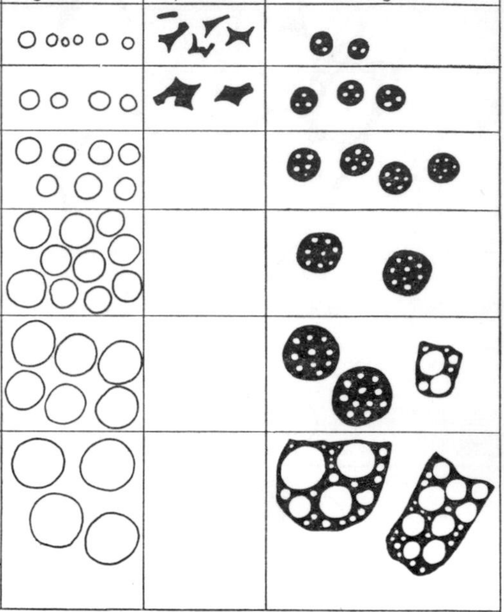

FIG. 25. Proportions and nature of particles of various sizes present in a flour commercially milled from English wheat. (Reproduced from Jones *et al.*, *J. Biochem. Microbiol. Technol. Engng.* **1**: 77, 1959, by courtesy of Interscience Publishers.)

suitable for the fine grinding of flour is a pinned disc grinder, consisting of a rapidly rotating disc studded with steel pins which intermesh with other pins on a similar but stationary disc (Jones *et al.*, 1959). Pinned disc grinders are suitable because they effectively break up the particles consisting of starch and protein without unduly damaging the starch granules themselves. Excessive starch damage (cf. p. 175) is to be avoided because it leads to increase in maltose value and deterioration in baking quality, and also because the damaged starch granules ("ghosts") enter the fine (0–15 μ) fraction and reduce its protein content (cf. p. 146).

Air classification. Sieving processes cannot generally be used for making separations below about 80 μ particle size. To obtain separations at 15 and 40 μ, the flour as milled, or after fine grinding, is fractionated by means of air classification, a process in which the effect of centrifugal force is opposed to the effect of air drag on individual particles. The cut size at which the separation is made can be adjusted, e.g. by altering the volume of air or its direction of travel. The data in Table 40 give an example of what can be accomplished by air classification of ground and unground flour from hard and soft wheats.

TABLE 40

YIELD AND PROTEIN CONTENT OF AIR-CLASSIFIED FRACTIONS OF FLOURS WITH OR WITHOUT PINNED DISC GRINDING*

Flour	Fine (0–17 μ)		Medium (17–35 μ)		Coarse (over 35 μ)	
	Yield (%)	Protein content[†] (%)	Yield (%)	Protein content[†] (%)	Yield (%)	Protein content[†] (%)
Hard wheat flour:						
Unground	1	17·1	9	9·9	90	13·8
Ground	12	18·9	41	10·0	47	14·7
Soft wheat flour:						
Unground	7	14·5	45	5·3	48	8·9
Ground	20	15·7	71	5·0	9	9·5

*Source: Kent (1965), with additional original data.
[†] 14% m.c. basis.

Fraction 1 can be used to increase the protein content of bread flours, particularly those milled from low or medium protein wheats, and in the manufacture of gluten-enriched bread and starch-reduced products. Fraction 2, when chlorinated (cf. p. 136), is recommended for sponge cakes and pre-mix flours. Fraction 3 is said to be of value for biscuit manufacture on account of its uniform particle size and granular structure.

The air classification process has two practical purposes:

1. It permits a range of flours, with varying properties and suitable for various bakery products, to be obtained from one parent flour. In a country, such as the U.S.A., in which particular types of wheat tend to be grown in localized areas (e.g. soft wheat in the Pacific Northwest, hard wheat in the Great Plains area), transportation of wheat for special purposes from other areas can be reduced or eliminated if flour, milled from local-grown wheat, can be adapted for other purposes by air classification.

2. It makes possible the production of flour of uniform characteristics by appropriate blending of air-classified fractions. The air classification process can assist in levelling off the variations in quality due to season and locality in an area, such as the southern Great Plains of the U.S.A., where wide variations in quality are encountered.

The maltose content of the fine air-classified fraction is higher than that of the original flour, because damaged starch granules ("ghosts") enter this fraction on account of their low terminal velocity. The fine fraction also has a relatively high water absorption on account of the damaged starch content; the high water absorption is advantageous, but the excessive diastatic activity is undesirable, and limits the use of this type of flour in the bakery.

Jones *et al.* (1960) found that the thiamine and nicotinic acid contents of the fine fraction tend to be higher than those of the parent flour on account of preferential entry of scutellum and aleurone fragments into the fine fraction; the riboflavin content of the fine fraction also tends to be higher, possibly because of the combination of this vitamin with the endosperm protein.

Preferential entry of fine mineral matter, aleurone and scutellum fragments into the fine fraction results in the colour of the fine fraction being duller than that of the parent flour, whereas the intermediate fraction tends to be brighter. The coarse fraction may also be duller than the parent flour in so far as small particles of bran present in the parent flour concentrate in the coarse fraction.

FLOUR FOR VARIOUS PURPOSES

Wheat flour is used for making food products of widely varying moisture content (cf. p. 177):

Level	Moisture content (%)	Type of food
High	90	Soup
High	45–70	Puddings
Medium	35	Bread, cakes (cookies)
Low	2	Biscuits (crackers)

For each purpose, flour of particular but varying properties is required; for breadmaking, a flour containing a high level of good quality protein; for biscuits, less protein is required, and the gluten should be extensible rather than elastic; for soups, the flour/water mixture should reach and maintain a certain viscosity when heated to 95°C. The quality of the starch and absence of amylolytic enzymes are of more importance than protein content for quality of soup flour.

In Britain, all types of flour, except wholemeal, must contain certain minimum levels of iron, vitamin B_1 and nicotinic acid (cf. p. 168), and all types of flour, except wholemeal and self-raising flour, must contain added chalk (calcium carbonate; cf. p. 164).

Bread (cf. Ch. 10)

The predominance of wheat flour for making aerated bread is due to the properties of its protein which, when the flour is mixed with water, forms an elastic substance called gluten (cf. pp. 43 and 173). This property is found to a slight extent in rye but not in other cereals.

The property of producing a loaf of relatively large volume, with regular, finely vesiculated crumb structure, is possessed by wheats described as "strong" (cf. pp. 68 and 179). Attempts have been made to associate good breadmaking quality with certain characteristics of strong wheats: it is positively correlated with protein content but dependent also on protein "quality".

When two flours of equal protein content produce loaves of different quality, the one flour is said to have protein of better quality than the other. Protein quality may be related to the proportion of the protein that is water-soluble (cf. p. 43).

Self-raising flour

Self-raising flour (self-rising flour, in the U.S.A.) is used for making puddings, cakes, pastry. It is milled from a grist consisting of weak wheats of low protein content, such as British or Australian, although up to 20% of strong wheat may be included.

The moisture content of the flour should not exceed 13·5% in order to·avoid premature reaction of the aerating chemicals and consequent loss of aerating power.

The maltose figure should be less than 3; hence, exclusion of sprouted wheat from the grist is important (cf. p. 87), as high diastatic activity leads to the production of dextrins and gummy substances during cooking, and to sticky and unattractive baked goods. Choice of sound wheat is also important because evolution of gas during baking is rapid and the dough must be sufficiently distensible, and yet strong enough to retain the gas.

Distension of the dough is caused by carbon dioxide which is evolved by the reaction of the raising agents (leavening agents), one alkaline and one acidic, in the presence of water. The usual agents are sodium bicarbonate and acid calcium phosphate (ACP), used at the rate of $3\frac{1}{4}$ lb of bicarbonate plus $4\frac{1}{2}$ lb of 80% grade ACP per 280 lb sack. A slight excess of the acidic component is desirable, as excess of bicarbonate gives rise to an unpleasant odour and a brownish-yellow discoloration.

Chalk is not added to self-raising flour, which generally has a calcium content of not less than 0·2% on account of the added ACP.

Household flour

This is similar to self-raising flour in quality, but without raising agents.

Biscuit (cracker) flours

There are many types of biscuit for which special types of flour are demanded. Biscuit flour is typically milled from weak wheat of low protein content, and many varieties of British wheat are ideally suited for this purpose. An all home-grown grist is generally used in Britain for milling biscuit flour.

The rheological properties of the biscuit dough are most important. A flour is required that makes a dough having more extensibility but less spring (resistance) than those of a bread dough. Spring, or recovery, is not required because the dough pieces should retain their size and shape unaltered after being stamped out. Any drawing together of the dough, reducing the diameter and increasing the thickness, would be a disadvantage, particularly for biscuits packed by machine, where exactness of dimensions of the finished article is most important.

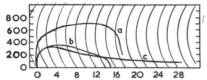

FIG. 26. Extensometer curves of unyeasted doughs made from biscuit flours milled from English wheats. *a*, poor biscuit flour—dough too elastic; *b*, poor biscuit flour—dough short; *c*, very good biscuit flour. (Unpublished curves by P. Halton.)

Measurements of dough spring or resistance (curve height S, R) and dough extensibility (curve length E) may be made by stretching unyeasted doughs on an extensometer which records on a chart the force required to stretch doughs and the extent to which the dough stretches before it breaks (cf. p. 153).[1] Halton, who has used this method of investigating biscuit flours, found that good biscuit flours had high extensibility and low spring ($E/S>9$), whereas flours unsatisfactory for biscuits had a low ratio ($E/S<7$) (see Fig. 26).

[1] The height of the curve was measured in Halton's original machine in terms of units of spring (S: maximum 10), but in the commercial model marketed by Henry Simon Ltd. it is measured in terms of resistance (R: maximum 1000).

The relationship between biscuitmaking quality and protein content of flour is not simple; although low protein wheat is generally specified for biscuit milling, an increase in protein content may sometimes accompany improvement in biscuitmaking quality. Thus, spring dressings with nitrochalk on English wheat increased protein content by 2·2% but also increased extensibility from 20 to 24 and decreased spring from 1·8 to 1·6, thereby increasing the ratio E/S from 11 to 15, and improving biscuitmaking quality.

The extensibility of biscuit-flour dough may be increased by treatment of the flour with a proteolytic enzyme, or with the reducing agent sulphur dioxide. Treatment of flour, other than wholemeal, intended for biscuit manufacture, with sulphur dioxide is permitted in Britain, but, from September 1964, the rate of addition may not exceed 200 ppm.

Flour for confectionery

Kent-Jones and Amos (1957) classified confectionery goods and their flour requirements as follows:

(a) Goods produced by fermentation: buns, etc. A breadmaking flour is required. Fermentation time is short; the fat and sugar in the formula bring about shortening of the gluten.

(b) Goods produced by chemical aeration: self-raising flour is used.

(c) Goods of sponge type: a weak, soft flour of fine particle size is required, such as "high-ratio" flour.

(d) Goods of puff or flaky type: strong baker's flour is required for puff pastry; weak, household flour for short pastry.

(e) Cakes: a medium strength flour is required for cakes containing a minimum of fruit. "High-ratio" flour—i.e., the finest fraction, dressed through a No. 16 or No. 20 silk (aperture length 0·09 or 0·07 mm), of the patent flour milled from a weak grist, and treated heavily with chlorine (e.g. 4 oz/sk) in order to modify the starch—is used for goods employing a high ratio of sugar and liquid to flour.

Flour for export

Besides the specific requirements according to the purpose for which the flour is to be used, flour for export must have low moisture content to prevent development of taint and infestation. As a safeguard, flour for export should be entoleted (cf. p. 141). In addition, export flour must conform to any special requirements ·of the importing country, e.g. regarding freedom from insect and rodent hair fragments for export to the U.S.A. (cf. p. 155).

Wholemeals (cf. pp. 129, 159 and 162)

Wholemeals (100% extraction) are made from a strong grist of good gluten quality. In the U.S.A., the 100% product is known as wholewheat flour or bread. Proprietary brown flours of 95–100% extraction, from which a small proportion of the coarsest bran has been removed, are also milled.

FLOUR TESTING

Further details of the tests mentioned below may be found in larger reference books, e.g. *Modern Cereal Chemistry, Cereal Laboratory Methods*.

Test baking

Flour test-baking procedures are chosen in relation to the ultimate purpose for which the flour is intended: viz. breadmaking, cakemaking, biscuitmaking tests are the final criteria by which the quality of bread, cake and biscuit flours, respectively, are judged. However, conditions under which such tests are carried out must be carefully standardized. Moreover, the interpretation of the results is not wholly objective. Hence, simpler, more objective tests have been sought, the results of which will correlate with, and forecast, the results of baking tests.

Protein test

The washing out of gluten was formerly used as a test for quality; the crude gluten can be dried and weighed. Although still widely used on the continent of Europe, this test has been superseded in Britain and the U.S.A. by estimation of nitrogen content, e.g. by the Kjeldahl method, and it is recognized that the nitrogen determination is the most important single test for flour quality. As 100 g of gluten protein contain on average 17·55 g of nitrogen, the gluten protein content of flour is calculated by multiplying the determined nitrogen content (percentage) by 5·7.

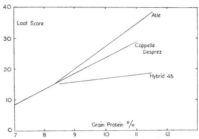

FIG. 27. Relationship between bread quality (expressed as loaf score) and grain protein content for three varieties of English-grown wheat. (Reproduced from E. N. Greer, *Milling* **137**: 560, 1961, by courtesy of the Northern Publ. Co. Ltd.)

A linear relationship between protein content and loaf volume has been found over the range 10·5–22·7% of protein, with coefficients of correlation between protein content and loaf volume as high as +0·99 *for samples of a given variety of wheat.* However, the slope of the line relating the two variables varies from variety to variety (see Fig. 27).

Rapid colorimetric methods of determining protein content include that of Udy, and the Pro-Meter method.

Sedimentation test

The sedimentation test, devised by Zeleny, is essentially a simplified water retention capacity test in the presence of lactic acid. The baking properties of the flour depend largely on the amount and quality of the gluten proteins, and the latter are caused to hydrate and swell by the lactic acid. Flour, water and lactic acid are shaken together in a glass cylinder under specified conditions, and the height of the sediment subsequently read. This test is said to measure both protein content and protein quality and hence to be superior to protein determination as a means of forecasting baking behaviour. Data from some hundreds of samples of HRS and HRW wheat from the U.S. 1962 harvest, examined by Zeleny and co-workers, were said to support this view, largely because of some high coefficients of correlation between sedimentation value and "flour evaluation score" (an index based largely on loaf characteristics) when comparisons were made within locality areas. However, critical assessment of the results of sedimentation, protein, and baking quality tests by workers in Britain and the U.S.A. (see, e.g., Schlesinger, 1963, for review) has shown that in general the sedimentation value is no better than the protein content for the prediction of baking quality, and that its limited agreement with the latter is largely a reflection of its dependence upon protein content. Moreover, the coefficient of partial correlation between sedimentation value and baking quality at constant protein content (a useful indication of the true worth of the sedimentation value as a measure of protein quality) is much too near zero to be acceptable in this respect. Nevertheless, the sedimentation test is being applied widely in the U.S.A. on account of its use in fixing price supports to farmers (cf. p. 76).

Physical tests on doughs and slurries

The *Brabender Farinograph* measures the plasticity and mobility of dough when subjected to continuous mixing at constant temperature.

The *Brabender Extensograph* and the *Research Extensometer* record the resistance of dough to stretching, and the distance the dough stretches before breaking (see Fig. 28).

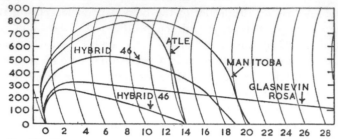

Fɪɢ. 28. Extensometer curves for unyeasted doughs made from the flour of various English-grown wheats in comparison with flour from Manitoba wheat. (Reproduced from P. Halton, *Soc. Chem. Ind. Monograph 6*, 1959, by courtesy of the Society of Chemical Industry.)

The *Chopin Alveograph* uses air pressure to inflate a bubble of dough and break it; the instrument continuously records the air pressure and the time that elapses before the dough breaks.

The *Chopin Zymotachegraph*, the *Brabender Fermentograph* and the Sandstedt & Kneen *Pressuremeter* measure gas production and gas retention by dough.

The *Brabender Amylograph* continuously measures the resistance to stirring of a 10% suspension of flour in water while the temperature of the suspension is raised at a constant rate of $1\frac{1}{2}°C/min$ from room temperature to 95°C and then maintained at 95°C. It is of use in testing flour for soups, etc., for which purpose the viscosity of the product after gelatinization is an important characteristic (cf. p. 147), and for adjusting the malt addition to flours for breadmaking.

The *Research Water Absorption Meter* measures the rate of extrusion of dough through a small orifice under controlled conditions. By testing doughs with varying water contents and graphing the results, the required quantity of water to produce doughs of standard flow characteristics can be estimated.

Tests for degree of milling refinement

Ash test. The ash test is widely used as a measure of milling refinement because the ash content of pure endosperm is relatively low whereas that of bran, aleurone and germ is relatively high (cf. p. 39). The ash test can be made very precisely.

Colour grade. Colour grade, estimated, e.g. with the Kent-Jones and Martin Colour Grader, can be used to estimate the degree of contamination of the flour with bran particles. The colour grade value is said to be unaffected by variation in content of yellow flour pigment (xanthophyll).

Moisture content. The moisture content of flour is a most important characteristic, particularly in relation to safe storage (cf. p. 90). The determination of moisture content is conveniently made by recording the weight loss when flour is heated at 100°C for 5 hr in vac., or at 130°C for 1 hr at atmospheric pressure. Electrical conductivity methods are also widely used; they are very rapid but less accurate than oven methods.

Maltose figure. This is used as an indication of (1) the amount of mechanical damage to starch caused by the milling process, and (2) the presence and amount of sprouted grain in the grist (cf. p. 174). The test procedure is described in standard textbooks (q.v.).

Extraneous matter test ("filth test"): cf. p. 151. The rodent hair and insect fragment count of flour is determined by digesting the flour with acid and adding the cooled digest to petrol in a separating funnel. The hair and insect fragments are trapped at the petrol/water interface, and can be collected and identified microscopically.

REFERENCES

AMERICAN ASSOCIATION OF CEREAL CHEMISTS (1962), *Cereal Laboratory Methods*, 7th edition, A.A.C.C. St. Paul, Minn., U.S.A.

BENTLEY, H. R., McDERMOTT, E. E., PACE, J., WHITEHEAD, J. K. and MORAN, T. (1950), Toxic factor from "Agenised" protein, *Nature, Lond.* **165**: 150. Action of nitrogen trichloride on proteins, *ibid.* 735.

BENTLEY, H. R., McDERMOTT, E. E., MORAN, T., PACE, J. and WHITEHEAD, J. K. (1950, 1951), Action of nitrogen trichloride on certain proteins, *Proc. Roy. Soc.* B **137**: 402; B **138**: 265.

HALTON, P. (1952), Wheats for biscuit flours, *Food Manuf.* **27**: 149.

HAWTHORN, J. and TODD, J. P. (1955), Some effects of oxygen on the mixing of bread doughs, *J. Sci. Fd. Agric.* **6**: 501.

HESS, K. (1955), Bedeutung für Teig, Brot und Gebäck von Zwickelprotein und Haftprotein in Weizenmehl, *Kolloid Z.* **141**: 61.

HLYNKA, I. (1964) (Ed.), *Wheat: Chemistry and Technology*, Amer. Assoc. Cereal Chem., St. Paul, Minn., U.S.A.

JONES, C. R., FRASER, J. R. and MORAN, T. (1960), Vitamin contents of air-classified high- and low-protein flour fractions, *Cereal Chem.* **37**: 9.

JONES, C. R., HALTON, P. and STEVENS, D. J. (1959), Separation of flour into fractions of different protein contents by means of air classification, *J. Biochem. Microbiol. Technol. Engng.* **1**: 77.

KENT, N. L. (1965), Effect of moisture content of wheat and flour on endosperm breakdown and protein displacement, *Cereal Chem.* **42**: 125.

KENT-JONES, D. W. and AMOS, A. J. (1957), *Modern Cereal Chemistry*, 5th edition, Northern Publ. Co. Ltd., Liverpool.

LOCKWOOD, J. F. (1960), *Flour Milling*, 4th edition, Northern Publ. Co. Ltd., Liverpool.

MELLANBY, E. (1946), Diet and canine hysteria, *Brit. Med. J.* **ii**: 885.

MINISTRY OF AGRICULTURE, FISHERIES AND FOOD (1963), *The Bread and Flour Regulations 1963*, Statutory Instruments 1963, No. 1435, H.M.S.O., London.

PACE, J. (1959), Chemical aspects of wheat proteins, in *The Physicochemical Properties of Proteins*, Chem. Ind. Monographs **6**: 26.

SCHLESINGER, J. S. (1963), Quality evaluation for the 1963 Oklahoma Hard Red Winter Wheat crop at harvest, *Northw. Miller* **269** (12): 10.

ZELENY, L. (1960), *Wheat Strength and the Sedimentation Test*, U.S. Dept. Agric. Marketing Service, Washington, D.C.

ZELENY, L., DOTY, J. M. and KIBLER, W. E. (1963), Sedimentation as a measure of wheat quality—1962 crop, *Northw. Miller* **268** (2): 19.

FURTHER READING

ANON., Radiation disinfestation of grain, *Milling* **139**: 450, 1962.

CORRESPONDENCE COURSE, *Milling*, 2 Nov. 1962; 21 Dec. 1962; 25 Jan. 1963.

HILL, E. G., Use of contact insecticides in milling premises, *Milling* **139**: 416, 1962.

HUTCHINSON, J. B., Steam treatment of wheat: a new type of flour, *Chem. & Ind.* 1084, 1963.

JOINER, R. R., New powdered agent for flour maturing [Azodicarbonamide], *Northw. Miller* **267** (12): 30, 1962.

JONES, C. R., Determination of degree of particle fineness, *Milling* **139**: 362; 388, 1962.

WHITBY, K. T., Particle sizing in the milling industry, *Cereal Sci. Today* **6**: 49, 1961.

CHAPTER 9

NUTRITIONAL ATTRIBUTES OF WHEAT, FLOUR, BREAD

CONSUMPTION OF FLOUR

Flour and its principal baked product, bread, are the cheapest and most important of our staple foods; yet, the consumption of flour, in all its various forms, has been decreasing in Britain since the end of the Second World War, and a similar trend is observed in European countries, the U.S.A., Canada and Australia (see Fig. 29). This fall in consumption is related to the rising standards of living in these countries, bringing the ability to buy more varied and expensive foods.

NUTRITIVE VALUE OF WHEAT

The nutrients in wheat are carbohydrate (mainly starch), protein, fat, vitamins and minerals. Wheat is regarded primarily as a source of carbohydrate because starch is the preponderating chemical constituent, whilst its valuable contribution of protein, vitamins—particularly those of the B group—and minerals is often overlooked. The National Food Survey 1963 revealed that flour and bread supplied on average, in Britain, in 1961, the percentages of the total intake of calories and particular nutrients as shown in Table 41.

Thus, flour and bread contribute to the national diet an even larger proportion of protein than of calories; the contribution of the other nutrients includes a proportion of additives (see below, p. 168).

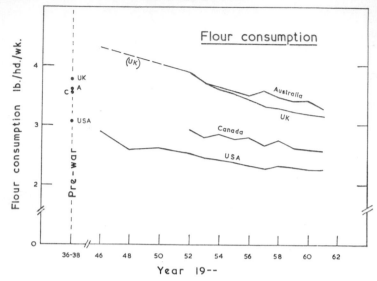

FIG. 29. Flour consumption, in pounds per head per week, in the U.K., Australia and Canada since 1952 and in the U.S.A. since 1946. Pre-war consumption (average for 1936–8) is shown by points on the dotted line. (From data in *Grain Crops*, 1959 and 1963; *Consumption of foods in the United States, 1909–1948*, Bureau of Agricultural Economics, U.S. Department of Agriculture, 1949.)

TABLE 41

PERCENTAGE OF ENERGY AND NUTRIENTS SUPPLIED
BY FLOUR AND BREAD IN BRITAIN IN 1961

Energy (calories)	21·0%
Protein	23·1%
Vitamin B_1	27·7%
Nicotinic acid	23·2%
Calcium	20·4%
Iron	24·8%

Protein (cf. p. 42)

Protein is the essential "body-building" nutrient, necessary for the growth, repair and maintenance of body tissues. The protein content of wheats varies over a wide range (6–21%) and is influenced less by heredity than by the edaphic factors—soil and climatic conditions—prevailing at the place of growth. Ranges of protein content encountered among samples of various wheat types are shown in Table 42.

TABLE 42
PROTEIN CONTENT OF WHEAT TYPES

Wheat type	Approx. protein range (%)*
Pusa (Australian)	13·7–15·3
Durum	10·3–16
Plate	10·0–15·8
Manitoba	10·0–15
United States HRS	10·5–12·8
United States HRW	9·6–14·8
Russian	9·0–14·7
Australian	8·0–13·6
Persian	10·0–12·2
English and Irish	6·8–12·5
Other European	7·9–11·6
United States SRW	8·8–11·1
Pacific (white)	8·0–11·5
Indian	8·7–10·8

* Air dry moisture basis.
Data for United States HRS, HRW and SRW from Shellenberger (1958); other data from Kent-Jones and Amos (1957).

Among samples of a particular type of wheat, the protein content is positively correlated with the vitamin B_1, nicotinic acid, iron and total mineral contents.

NUTRITIVE VALUE OF FLOUR AND BREAD

The nutritive value of 100% wholemeal is the same as that of wheat, because wholemeal must (in Britain) contain the whole of the products derived from the milling of clean wheat (Bread and Flour Regulations 1963), but flours of lower extraction rates differ from wheat in nutritive value because of the removal of varying amounts of bran, germ and outer endosperm, containing higher concentrations of protein, minerals and vitamins.

Extraction rate

The weight of flour produced per 100 parts of wheat milled is known as the flour yield or percentage extraction rate. In practice, flour yield is generally calculated in Britain as a percentage of the products of the milling of clean wheat (cf. p. 165), whereas in the U.S.A. yield is expressed in terms of the number of bushels and pounds of clean wheat required to produce 100 lb of flour.

The wheat grain contains about 82% of the white starchy endosperm which is required for white flour (cf. Table 9), but it is never possible to separate it exactly from the 18% of bran, aleurone and germ and thereby obtain a white flour of 82% extraction. The mechanical limitations of the milling process are such that in practice 75% is about the limit of white flour extraction, further increase darkening the colour through inclusion of a proportion of bran, aleurone and germ. The following figures for proportions of bran, germ and endosperm in flours of various extraction rates are quoted by Moran and Drummond (1945):

	Rate of extraction (%)		
	100	85	80
Bran	12	3·4	1·4
Germ	2·5	1·9	1·6
Endosperm	85·5	79·7	77·0

White flour. In Britain, between the world wars, flour-millers were free to mill flour to any extraction rate, and chose the rate that gave them the best financial return. When extraction rates are thus freely chosen, the price flour fetches is dependent to some extent on flour colour, and colour becomes poorer as extraction rises beyond a certain limit which varies a little from mill to mill. Thus, when a miller makes a flour of longer extraction rate, he has more of it to sell than when he makes shorter extraction flour, but he may get a poorer price for it or may not find a market at all. Reduction in extraction rate below 70% causes very little change in flour colour, so the optimum extraction rate, prevailing until 1939 in Britain, was about 70%.

White rather than brown flour is milled because of public preference for white bread. With both white and brown breads available, sales of brown have never exceeded 5–10% of the total since 1914; moreover, white flour always commands a better price than brown on a free market. Furthermore, white flour without the germ and bran keeps better and does not become rancid or infested with insects so readily as brown flour does.

It is true that the germ and aleurone (attached to the bran), rejected as part of the offals in the milling of white flour, contain a concentration of valuable vitamins and minerals; however, the loss of the most important of these nutrients is made good by enrichment of white flour with vitamin B_1, nicotinic acid and iron (cf. p. 168). Moreover, the bran in the brown flour is poorly digested by humans and is better utilized in the form of wheat-feed by cows, which provide milk in return.

Higher extraction flours. Flours of extraction rates between 75% and 100% can be milled by adding to white flour a proportion of the reground offals from 75% extraction milling until the desired extraction rate is reached (cf. p. 126–129). A series of flours, all of the same extraction rate, can be made by adding the appropriate quantity of all or any of the available offal streams, and such flours will show a wide range in composition. Thus, the extraction rate of a flour is not necessarily a guide to its composition unless its method of manufacture is specified in some way, e.g. by the addition of successive increments of the lowest ash content flour available, and even then composition of flours of equivalent extraction rate will vary from mill to mill.

If a series of flours of varying extraction rates is prepared in a particular mill from the same grist, keeping the bran content as low as possible at any extraction rate, the contents of nearly all the nutrients will increase as extraction rate rises, but the rate of increase is non-linear and is characteristic for each nutrient— see Fig. 30. The non-uniform distribution of the various chemical constituents among and within the various parts of the grain accounts for the particular shape of the curves in Fig. 30; e.g. the protein concentration in the endosperm increases from the centre to the periphery, and is particularly high in the aleurone and germ (cf. pp. 44–45).

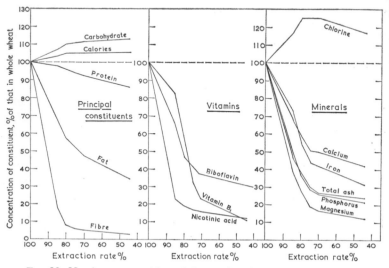

FIG. 30. Nutrient composition of flours of various extraction rates in relation to that of whole wheat. (Constructed from data in McCance *et al.*, *Biochem. J.* 39: 213, 1945.)

Flours of increasing extraction rate contain less carbohydrate and chlorides but more of all other nutrients. Carbohydrates and chlorides must therefore be concentrated in the inner endosperm, whereas other nutrients are concentrated elsewhere. Increasing proportions of germ tend to be included in the main product when extraction rate rises from 75 to 85%, and hence the curve for vitamin B_1, which is concentrated mainly in the scutellum part of the germ, rises steeply between these extractions, but flattens out at higher extraction rates. Increasing proportions of bran are included at extraction rates between 85 and 100%, and hence curves for fibre (mainly in the pericarp) and nicotinic acid (concentrated in the aleurone layer) tend to be flat at extraction rates below 85%, but rise steeply at extraction rates between 85 and 100%.

Flour of 85% extraction rate has a higher content of all nutrients (except carbohydrate) than 75%, but contains considerably less indigestible fibre than 100% wholemeal. Wheatmeal or brown flour is required by British Government regulation (Bread and Flour Regulations 1963) to contain not less than 0·6% of fibre (d.m.b.).

Wheatfeed (millfeed, in the U.S.A.). The yield of bran and fine wheatfeed (shorts, in the U.S.A.) made at any extraction rate is complementary to the yield of flour; the chemical composition of these by-products varies with extraction rate, as shown by data in Table 43.

TABLE 43
COMPOSITION OF FLOUR AND MILLING OFFALS AT VARIOUS
EXTRACTION RATES*

Material	Yield (%)	Protein (%)	Oil (%)	Ash (%)	Fibre (%)	Vitamin B_1 ($\mu g/g$)	Nicotinic acid ($\mu g/g$)
Flour:							
85% extraction	85	12·5	1·5	0·92	0·33	3·42	—
80% extraction	80·5	12·0	1·4	0·72	0·20	2·67	19
70% extraction	70	11·4	1·2	0·44	0·10	0·7	10
Fine wheatfeed (shorts):							
85% extraction	10	12·6	4·7	5·1	10·6	6·0	—
80% extraction	12·5	14·3	4·7	4·7	8·4	10·4	191
70% extraction	20	15·4	4·7	3·5	5·2	14·0	113
Bran:							
85% extraction	5	11·1	3·7	6·1	13·5	4·6	—
80% extraction	7	12·4	3·9	5·9	11·1	5·0	302
70% extraction	10	13·0	3·5	5·1	8·9	6·0	232

* From Jones (1958).

Phytic acid. Associated with the fibre of cereal grains is a substance, phytic acid (inositol hexaphosphate), which forms an insoluble compound with calcium and iron. More than 90% of the total phytic acid of wheat is localized in the aleurone layer (Pringle, 1952). High extraction flour, containing bran and aleurone, and therefore phytic acid, tends to immobilize the calcium and iron present in the flour and in other ingredients of a diet containing high extraction flour. At the beginning of the Second World War, when an increase in flour extraction rate was being contemplated, McCance and Widdowson (1942) reported: "To change a nation's dietary from white to brown bread, and at the same time to reduce their milk and cheese supply (the main sources of calcium) would probably mean that nine adults out of ten would begin to lose calcium. Rickets might increase in young children and growth become slower at all ages." This warning received confirmation in Eire in 1943 when, after three years of 100% wholemeal in Dublin, the incidence of rickets in young children had increased from a practically negligible percentage to 50%. Vitamin deficiencies (e.g. of vitamin D) may have played some part in these events, but the Irish investigators concluded that the high incidence of rickets in Dublin children during 1942–3 was probably caused by the high phytate content of the 100% extraction flour then in general use.

In order to counteract the effects of phytic acid in long extraction flour, the British Government ordered the addition of calcium carbonate (creta praeparata, or chalk B.P.) to 85% National flour at a rate of 156 mg per 100 g, in 1942, when the flour extraction rate was first increased to 85%. The rate of addition was doubled in 1946. It is now compulsory in Britain to add chalk (not less than 235 mg nor more than 390 mg per 100 g) to all flours except 100% wholemeal. If this exception might seem illogical (since wholemeal, of all types of flour, contains the largest amount of phytic acid, and would seem to require the largest addition of chalk), it must be remembered that consumers of this particular product are concerned to an exceptional extent with the concept of absence of all additions.

Strontium 90. Sr^{90} is a radioactive element which occurs as a by-product in the fission of the atoms of heavy metals, such as uranium, and thus forms part of the radioactive fall-out from atomic explosions.

Sr^{90} is important because, being chemically similar to calcium, it can replace calcium, e.g. in bones, causing irritation and disease. The human body takes up calcium four times as readily as Sr^{90}: thus, absorption of Sr^{90} is minimized in a diet rich in calcium. Flour correctly fortified with chalk provides about one-quarter of the total calcium in the diet, whereas its contribution to Sr^{90} intake is only about 12%. From this point of view, addition of chalk to white flour makes it probably the most useful main food item. The Sr^{90} contents of flour and bran have been reported as 1–5 and 10–81 $\mu\mu c$/kg, respectively (1 curie (c) Sr^{90} is about 0·02 g; 1 $\mu\mu c$ is thus about 2×10^{-14} g).

Phytase. Phytic acid is hydrolysed to phosphoric acid and inositol by the enzyme phytase, optimum activity occurring at 55°C. Probably 60% of the phytic acid in flour is hydrolysed during breadmaking.

National flour. From the outbreak of war in September 1939 until August 1953 the British flour-milling industry was controlled by the government. The extraction rate was fixed and, in order to conserve wheat supplies, was raised to 85% in March 1942. The 85% was called National flour, and a similar flour, varying in extraction rate between 80 and 90%, continued to be milled until decontrol in 1953. From 1942 until 1953, extraction rates were calculated with respect to weight of "dirty wheat" (actually, total products of milling including dust and screenings —cf. p. 160). Raising of the extraction rate, besides conserving wheat supplies, produced a flour of higher nutritive value.

Throughout 1943, in Britain, the wheaten grist was diluted with up to 10% of rye, barley or oats; in 1943 and 1944 a small amount of skim milk powder was added to National flour. A summary of the principal changes in National flour during this period is shown in Fig. 31.

High vitamin white flour. In 1940 the Medical Research Council (M.R.C.) recommended that 85% National flour should be milled so as to contain: (1) as much of the vitamin B_1, riboflavin

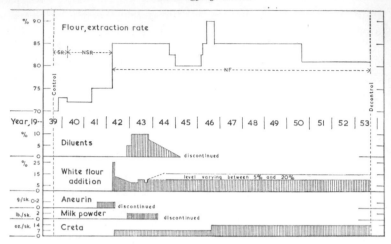

Fig. 31. The principal changes in National flour during the period of control by the British Government, September 1939 to August 1953. SR—straight run; NSR—National straight run; NF—National flour. For further explanation see text.

and nicotinic acid as possible, (2) as much protein as possible, particularly that derived from the outer part of the endosperm, and (3) as little of the bran as possible.

Flour-millers were assisted in implementing these M.R.C. recommendations by the results of Hinton's work, which had shown that most of the vitamin B_1 of the grain is located in the scutellum part of the germ. The appearance and nutritive value of 85% National flour was greatly improved, the fibre content progressively decreased, while the vitamin B_1 content was maintained. Eventually, an 80% flour was milled which was very nearly white and yet contained much more of the B vitamins than were present in white flour.

A similar flour milled in Canada by an extension of the ordinary milling process was known as "Canada Approved".

Enrichment. In 1938 the British flour-milling industry, on the advice of Prof. E. C. (now Sir Charles) Dodds, proposed the addition of synthetic vitamin B_1 (aneurin, thiamine) to white flour in order to raise the vitamin B_1 level to 0·24 mg per 100 g of flour. The government started enriching flour with vitamin B_1 in 1940, but discontinued the process in 1942, when 85% extraction flour became compulsory and enrichment with vitamin B_1 unnecessary (see Fig. 31).

In the U.S.A., enrichment of white flour and bread was started partly as a result of the economic depression of the 1930's. The National Research Council had recommended that thiamine (vitamin B_1), niacin (nicotinic acid), riboflavin and iron—four nutrients conspicuously lacking in the diet—could be restored and made available to all income groups by adding to white flour.

Deficiency diseases are caused by insufficiency of these nutrients in the diet: insufficiency of thiamine leads to beriberi; of riboflavin to impaired growth, dermatitis, eye defects; of niacin to pellagra; of iron to anaemia.

Synthetic thiamine and niacin were available commercially in 1940, and in 1941 U.S. millers were adding these and iron voluntarily to flour and bread. Synthetic riboflavin became available commercially in 1943, and in that year enrichment with all four nutrients became compulsory. In 1946 enrichment again became voluntary except in those states whose law requires it (now 29 states).

White flour or bread in the U.S.A. is said to be enriched if it is supplemented so that it contains nutrients between the levels shown in Table 44.

TABLE 44
ENRICHMENT STANDARDS IN THE U.S.A.

Constituent	Flour standards		Bread standards	
	Min.	Max.	Min.	Max.
Thiamine, mg/lb	2·0	2·5	1·1	1·8
Riboflavin, mg/lb	1·2	1·5	0·7	1·6
Niacin, mg/lb	16·0	20·0	10·0	15·0
Iron, mg/lb	13·0	16·5	8·0	12·5
Optional:				
Calcium, mg/lb	500	1500	300	800
Vitamin D, USP units	250	1000	150	750

In Britain, since 1953, the production of white flour has been permitted without stipulation as to extraction rate, provided that the content of certain nutrients is not less than that required for National flour which, in so far as any public demand for it persists, must be of 80% extraction.

The nutrients referred to were those recommended to be added to flour by the conference on the post-war loaf, 1945, namely,

	mg per 100 g of flour
Vitamin B_1 (thiamine)	0·24
Nicotinic acid (niacin)	1·60
Iron	1·65

These nutrient standards could be attained in flour of 80% extraction rate.

The Bread and Flour Regulations 1963 require all flour in Britain to contain these quantities of nutrients. In the case of flour described as wholemeal, such nutrients shall be naturally present and not added. In the case of flour not so described, such nutrients shall be added, where addition is necessary. In the case of iron, the addition is to be in the form of ferric ammonium citrate or of reduced iron.

The Post-war Loaf Conference also recommended that comparison be made between low extraction flours suitably reinforced with "token" nutrients (vitamin B_1, nicotinic acid and iron) and high extraction flour milled wholly from the wheat grain; the Medical Research Council arranged suitable trials in 1947 and 1948, which were carried out by Widdowson and McCance on children in orphanages in Germany. The results showed that when bread supplied 75% of the calories in a diet otherwise poor but containing plenty of vegetables, no significant difference could be shown between children fed on bread made from 70% flour, 85% flour, 100% wholemeal, or 70% flour enriched with vitamin B_1, riboflavin, nicotinic acid and iron either to the 85% or to the 100% level of these nutrients. The children, previously underweight and underheight, rapidly caught up with American standards for the best-fed children, when fed on any of these breads.

In view of the result of the Widdowson–McCance German experiments, the British Government would have been justified in advocating that a return to white flour enriched to the 80% level by addition of the three token nutrients (vitamin B_1, nicotinic acid and iron) would not be detrimental to the nation's health. However, in 1955, the Government referred the question to an independent panel under the chairmanship of Lord Cohen in an attempt to resolve it once and for all. The panel, which considered evidence from the baking and milling industries, the medical profession, government departments, and indirectly interested bodies and individuals, concluded, in their 1956 report, that the health of the nation could be safeguarded equally well by an 80% extraction flour or a white flour enriched with vitamin B_1, nicotinic acid and iron as prescribed by the Flour Order, 1953. This conclusion was reached by considering bread as part of a mixed diet and not merely by comparing the various flours and breads.

NUTRITIVE VALUE OF BREAD PROTEIN

It has been known since the classical work of Osborne and Mendel (1914) that wheat protein is deficient in the amino acid lysine (cf. Table 19). The work of Rose *et al.* (1954) has established that the requirement for lysine is higher than for any of the other essential amino acids. Thus, the nutritive value of wheat protein is limited by its low lysine content.

When protein is made the limiting factor, the growth rate of rapidly growing mammals (e.g. weanling rats) is slightly higher on wholemeal bread than on white bread, probably because of the slightly higher lysine content of the former.

Hutchinson *et al.* (1959) found, in feeding trials with young rats, that the growth rate could be increased nearly three-fold by addition of about 0·25% of L-lysine to a white bread diet. Further addition of lysine produced no further improvement of growth but, with a diet of protein content 12·8%, and a lysine supplement of 0·5%, there was a further response in growth if 0·1% of L-threonine (another amino acid) was also added. The rate of growth obtained on white bread so supplemented with L-lysine and L-threonine was about equal to the growth rate on a comparable carbohydrate–casein diet, although the growth rate on a comparable diet with whole egg protein was somewhat superior to that on any of these supplemented diets.

McDermott and Pace (1957) have shown that when flour is baked into bread there is a slight fall in the phenylalanine and tyrosine contents of the protein hydrolysates. The losses were somewhat higher from bread crust than from bread crumb, and there was also a small loss of lysine from the crumb hydrolysates. This apparently was due to reaction of the amino acids with sugars in the browning reaction.

Coeliac disease. Coeliac disease is a relatively uncommon condition of gluten intolerance in children. To such patients, wheat and rye are deleterious but other cereals are not. The symptoms are the passage of abnormal stools, abdominal distension, loss of weight, anaemia, anorexia and stunted growth (Frazer, 1962). Patients recover, or are improved, on a strict gluten-free diet.

The preparation of gluten-free bread from maize starch and non-cereal ingredients has been described by Zentner (1963).

Incubation of gluten with the mucous membrane from the small intestine of normal (gluten-tolerant) people destroys the deleterious agent in the gluten, but this does not happen if the mucous membrane is taken from a coeliac or gluten-intolerant patient (Frazer, 1962). Suggested explanations of the disease mechanism are deficiency of a specific enzyme in the intestinal mucosa (Weijers and van de Kamer, 1959), or reduced efficiency of the mucosa as a barrier that normally prevents absorption of the deleterious agents (Frazer, 1962).

REFERENCES

FRAZER, A. C. (1962), The possible role of dietary factors in the aetiology and pathogenesis of sprue, coeliac disease and idiopathic steatorrhoea, *Proc. Nutr. Soc.* **21**: 42.

HUTCHINSON, J. B., MORAN, T. and PACE, J. (1959), The nutritive value of bread protein as influenced by the level of protein intake, the level of supplementation with L-lysine and L-threonine, and the addition of egg and milk proteins, *Brit. J. Nutr.* **13**: 151.

JONES, C. R. (1958), The essentials of the flour-milling process, *Proc. Nutr. Soc.* **17**: 7.

KENT-JONES, D. W. and AMOS, A. J. (1957), *Modern Cereal Chemistry*, 5th edition, Northern Publ. Co. Ltd., Liverpool.

McCANCE, R. A. and WIDDOWSON, E. M. (1942), Mineral metabolism of healthy adults on white and brown bread dietaries, *J. Physiol.* **101**: 44.

McDERMOTT, E. E. and PACE, J. (1957), The content of amino acids in white flour and bread, *Brit. J. Nutr.* **11**: 446.

MEDICAL RESEARCH COUNCIL ACCESSORY FOOD FACTORS COMMITTEE (1940), MRC Memorandum on Bread, *Lancet* **ii**: 143; *Brit. Med. J.* **ii**: 164.

MINISTRY OF AGRICULTURE, FISHERIES AND FOOD (1963), *Domestic Food Consumption and Expenditure: 1961*, Annual Report of the National Food Survey Committee, H.M.S.O., London.

MINISTRY OF AGRICULTURE, FISHERIES AND FOOD (1963), *The Bread and Flour Regulations 1963*, Statutory Instruments 1963, No. 1435, H.M.S.O., London.

MINISTRY OF FOOD (1945), *Report of the Conference on the Post-war Loaf*, Cmd. 6701, H.M.S.O., London.

MINISTRY OF FOOD (1953), *The Flour Order, 1953*, Statutory Instruments 1953, No. 1282, H.M.S.O., London.

MORAN, T. and DRUMMOND, J. C. (1945), Scientific basis of 80 per cent extraction flour, *Lancet* **i**: 698.

OSBORNE, T. B. and MENDEL, L. B. (1914), Amino acids in nutrition and growth, *J. Biol. Chem.* **17**: 325.

PRINGLE, W. J. S. (1952), Mineral constituents of wheat and flour, in BATE-SMITH, E. C. and MORRIS, T. N. (1952).

Report of the Panel on Composition and Nutritive Value of Flour (1956) (Cohen Report), Cmd. 9757, H.M.S.O., London.

ROSE, W. C., HAINES, W. J. and WARNER, D. T. (1954), The amino acid requirements of man. V. The role of lysine, arginine and tryptophan, *J. Biol. Chem.* **206**: 421.

SHELLENBERGER, J. A. (1958), Survey of quality of European wheat imports, *Kansas Agric. Exp. Sta. Bull.* 396.

WEIJERS, H. A. and VAN DE KAMER, J. H. (1959), Coeliac disease. 7. Application and interpretation of the gliadine tolerance curve, *Acta Paediat.* **48**: 17.

WIDDOWSON, E. M. and McCANCE, R. A. (1954), *Studies on the Nutritive Value of Bread and on the Effect of Variations in the Extraction Rate of Flour on the Growth of Undernourished Children*, Med. Res. Coun., Spec. Rep. Ser. 287, H.M.S.O., London.

ZENTNER, H. (1963), Coeliac disease: Biochemical and technological aspects, *Food Technol. Australia* **15**: 126.

FURTHER READING

JONES, C. R. and MORAN, T., A study of the mill streams composing 80% extraction flour with particular reference to their nutritional composition, *Cereal Chem.* **23**: 248, 1946.

MINISTRY OF AGRICULTURE, FISHERIES AND FOOD, *Food Standards Committee Report on Bread and Flour*, H.M.S.O., London, 1960.

MINISTRY OF FOOD, National flour and bread, *Nature, Lond.* **149**: 460, 1942; **150**: 538, 1942; **151**: 629, 1943; **153**: 154, 1944; **154**: 582, 788, 1944; **155**: 717, 1945; **157**: 181, 1946.

MORAN, T., Nutritional significance of recent work on wheat, flour and bread, *Nutr. Abs. Revs.* **29**: 1, 1959.

CHAPTER 10

TECHNOLOGY OF BAKING

PRINCIPLES

The first requirement in baking is aeration of the mixture by incorporation of a gas; the second is coagulation of the material by heating it in the oven so that the gas is retained, and the structure of the material is stabilized. The advantage of having an aerated, finely vesiculated crumb in the baked product is that it is easily masticated.

There are three stages in the manufacture of bread: mixing the dough, dough fermentation, and oven baking of the dough.

The main ingredients of the dough are wheaten flour, water, yeast and salt. Other ingredients which may be added include malt flour, yeast foods, milk and milk products, fat, fruit, gluten.

When these ingredients are mixed in correct proportions to make a dough, two processes commence: (1) the protein in the flour begins to hydrate, i.e. to combine with some of the water, to form a material called gluten (cf. pp. 43 and 147), which has peculiar extensible properties—it can be stretched like elastic, and possesses a certain degree of recoil or spring; (2) evolution of the gas carbon dioxide by action of the enzymes in the yeast upon the sugars.

Enzymes. The enzymes principally concerned in panary fermentation are those that act upon carbohydrates: alpha-amylase and beta-amylase in flour, often spoken of together as diastase, and maltase, invertase and the zymase complex in yeast. Zymase is an old collective name for about fourteen enzymes.

The starch of the flour is broken down to the disaccharide maltose by the amylase enzymes; the maltose is split to glucose (dextrose) by the maltase; cane sugar present in the flour is split to glucose and fructose (laevulose) by invertase; and glucose and fructose are fermented to carbon dioxide and alcohol by the zymase complex.

Amylase. Both alpha- and beta-amylase attack starch, but in different ways. Alpha-amylase attacks α–1 : 4 (glucosidic) linkages (cf. p. 40) of amylose and amylopectin at almost any point in the molecule, liberating portions with a non-reducing end, but is unable to attack the 1 : 6 linkages at the branches of the amylopectin molecule. The residue from alpha-amylase attack on starch is dextrins of low molecular weight, and this enzyme is thus sometimes called the dextrinogenic amylase.

Beta-amylase also attacks α–1 : 4 linkages of amylose and amylopectin, but can do so only from the non-reducing end of the molecule, splitting off two glucose units at a time, in the form of maltose, until blocked by the proximity of a linkage other than 1 : 4 (e.g. the 1 : 6 linkages in amylopectin). The beta-amylase is also called the saccharogenic amylase. In the absence of alpha-amylase, beta-amylase breaks down about one-third of the amylopectin, leaving a highly resistant residue of higher molecular weight dextrins, known as "limit dextrin". However, in the presence of minute amounts of alpha-amylase, the beta-amylase is able to renew its attack at the non-reducing ends of the portions of the molecules liberated by alpha-amylase attack.

Normal flour from sound wheat contains ample beta-amylase but generally only a small amount of alpha-amylase. The amount of alpha-amylase, however, increases considerably when wheat germinates. Indeed, flour made from wheat containing many sprouted grains may have too high an alpha-amylase activity, with the result that, during baking, sufficient of the starch is changed into dextrin-like substances to weaken the crumb and make it sticky (cf. p. 148).

The amount of alpha-amylase in a flour which has inadequate dextrinogenic activity may be increased by addition of malt or of fungal amylases.

Water absorption. When water and flour are mixed in dough-making the water is absorbed largely by the protein and starch. The amount of water absorbed to make a dough of standard consistency increases in proportion to the contents of protein and damaged starch present. For use in mechanical bakeries it is particularly important that the water absorption of the flour be maintained at a uniform level, as adjustment of the amount of water necessary to make dough of standard consistency is more inconvenient there than in small-scale baking.

Starch. Some of the starch granules in flour become damaged during the flour-milling process. The damaged granules may be recognized, microscopically, by a red coloration with Congo Red stain (cf. p. 221), and have been described, from their microscopical appearance, as "ghosts" by Jones (1940). The undamaged granules do not stain with Congo Red.

Two factors causing starch damage have been recognized by Jones (1940): a surface factor, depending on the degree of shearing or scraping of the material during milling, and related to the roller surface characteristics and the speed differential between the two rolls (cf. p. 125); and an internal factor, depending on the degree of flattening and crushing of the endosperm particles, and related to roll pressure.

The proportion of granules damaged depends on the severity of milling and on the hardness of the endosperm, and averages about 9% in bread flours. More damage is sustained by the flour of hard wheat, which requires heavier rolling to reduce it to flour fineness than soft wheat does. It is believed that flour amylases are able to attack only the damaged or "available" starch. It is therefore essential that the flour contain adequate damaged starch to supply sugar during fermentation. Severe overgrinding during milling, causing excessive starch damage, has an adverse effect on bread quality: water absorption is increased, bread volume is decreased, and the bread is less attractive in appearance (cf. p. 144).

Sugar. There are small quantities of sugar naturally present in the flour (cf. p. 41), but these are soon used up by the yeast, which then depends on the sugar produced by diastatic action from the starch.

The brown colour of the crust of bread is probably due to a non-enzymic browning reaction (Maillard type) in which protein reacts with reducing carbohydrates (Bertram, 1953). The glaze on the crust is due, in part, to starch gelatinization which occurs when the humidity is high.

Fermentation. Glucose and fructose are fermented to carbon dioxide (which aerates the dough) and alcohol in a number of stages by a large group of enzymes and coenzymes formerly known as the zymase complex. The first stage is a phosphorylation of the hexose sugars to esters of phosphoric acid; intermediate products of later stages include phosphoglyceraldehyde, phosphoglyceric acid, pyruvic acid and acetaldehyde (see Pyler, 1952, for further details). Some of the coenzymes involved in these reactions are of interest because they contain vitamins as components of their structure: cozymase contains niacin or nicotinamide; cocarboxylase contains thiamine.

About half a proof gallon of alcohol is produced per 280 lb sack of flour during fermentation, but most of it is driven off during the baking process. New bread is said to contain up to 0·3% of alcohol. Secondary products, e.g. acids, carbonyls and esters, may affect the gluten, or may impart flavour to the bread.

Gas production and gas retention. Adequate but not excessive gas must be produced during fermentation, otherwise the loaf will not be inflated sufficiently. Gas production depends on the quantity of soluble sugars in the flour, and on its diastatic power and granulation. Flours with high maltose values (e.g. above 3 by Blish and Sandstedt method), due to excessive starch damage through overgrinding, or to high alpha-amylase activity, may not bake satisfactorily. In the former case, the overgrinding may also damage the gluten, giving short doughs; in the latter case, the starch, after enzymic attack, fails to hold water adequately at baking, giving poor baking quality (cf. p. 148). Inadequate gassing (maltose values less than 1·5) is less serious, and can be corrected by adding sprouted wheat to the grist, or malt flour to the flour.

Gas retention is a property of the flour protein: the gluten, while being sufficiently extensible to allow the loaf to rise, must yet be strong enough to prevent gas escaping too readily, as this would lead to collapse of the loaf.

Proteolytic enzymes. Besides the enzymes that act on carbohydrates, there are many other enzymes in flour and yeast, of which those that affect protein, the proteolytic enzymes, may be of importance in baking. Yeast contains such enzymes, but they remain within the yeast cells and hence do not influence the gluten.

The proteolytic enzymes of flour are proteases. It can be shown from study of flour–water doughs that amino-nitrogen is liberated from flour by enzyme action during fermentation, but, in yeasted doughs, the release of amino-nitrogen is less obvious because the latter is quickly used up by the yeast.

The relationship between flour proteases and oxidizing substances that act as improvers (cf. p. 138) and their importance in dough "ripening" is a debated subject. The effect of proteases may be disaggregating rather than proteolytic. Moreover, the undesirable effect of flour milled from wheat attacked by bug (cf. p. 86) is generally considered to be due to excessive proteolytic activity. Heat treatment inactivates proteolytic enzymes more rapidly than diastatic enzymes, and has been recommended as a remedy for this condition. However, it is difficult to inactivate enzymes by heat treatment without damaging the gluten proteins simultaneously.

Flour–water ratio for baked goods. The number of parts of water per 100 parts of flour used in baking depends on the type of product being made (cf. p. 147), and may vary from 20 for biscuit dough to 150 for wafer batter. For bread dough, 53–57 parts of water (viz. 15–16 gal per 280 lb sk of flour) are usual. The biscuit dough is stiff to permit rolling and flattening; the bread dough is a plastic mass that can be moulded and shaped; the wafer batter is a liquid suspension that will flow through a pipe.

BREADMAKING PROCESSES

Good bread quality implies that the loaf has sufficient volume, an attractive appearance as regards shape and colour, and a crumb that is finely and evenly vesiculated and soft enough for easy mastication, yet firm enough to permit thin slicing. The attainment of good quality in bread depends partly on the inherent characteristics of the ingredients—particularly the flour—and partly on the baking process. The technique employed in baking is related to the quantity and quality of the gluten formed from the flour protein and to the time taken for optimum gluten ripening. Bread is made by mixing a dough from flour, water, yeast and salt, allowing the dough to rest at a temperature of about 80°F—while fermentation and gluten ripening take place—and then baking in the oven.

Dough ripening. A dough undergoing fermentation, with inter-mittent mechanical action, is said to be ripening. The dough when mixed is sticky, but, as ripening proceeds, it becomes less sticky and more rubbery and is more easily handled on the plant, and the bread baked from it becomes progressively better, until an optimum condition of ripeness has been reached. If ripening is allowed to proceed beyond this point a deterioration sets in, the dough gets short and possibly sticky again, and bread quality becomes poorer. A dough ripe for moulding has maximum spring and elasticity; a green or under-ripe dough can be stretched but has insufficient spring; an over-ripe dough tends to break when stretched.

If the optimum condition of ripeness persists over a reasonable period of time the flour is said to have good fermentation toler-ance, but if the fermentation time giving optimum ripening is sharply defined, the flour has poor fermentation tolerance. Weak flours quickly reach a relatively poor optimum, and have poor tolerance, whereas strong flours give a higher optimum, take longer to reach it, and have good tolerance.

Flour. For breadmaking, flour from a grist containing a large proportion of strong wheat is required (cf. p. 68). Good bread-making flour is characterized by having protein which is adequate in quantity and of satisfactory quality in respect of elasticity, strength and stability, satisfactory gassing properties and amylase activity, satisfactory moisture content—not higher than about 14% to permit safe storage—and satisfactory colour.

The degree of strength required in breadmaking flour varies according to the baking system: a stronger flour is used for a sponge and dough process than for a straight dough process (*vide infra*).

Water. Flour from strong wheat requires more water than flour from soft wheat (cf. p. 175), and determination of the absorbing power, or "water absorption", of flour is important (cf. p. 154).

Yeast. Bakers' yeast is a different strain from brewers' yeast, and it must be fresh and active. The quantity used is related inversely to the time of fermentation and to the temperature of the dough. Longer fermentation systems generally employ somewhat lower dough temperatures. Thus, $3\frac{1}{2}$ lb of yeast per 280 lb sk of flour would be used for a 3 hr system with the dough at 80°F, whereas only $1\frac{1}{4}$ lb of yeast per sk would be required for an 8 hr system with the dough at 76°F.

Salt. Salt is added to develop flavour. It also toughens the gluten and gives a less sticky dough. Salt slows down the rate of fermentation, and thus more is used in longer fermentation systems than in shorter ones. The quantity added is $4\frac{1}{2}$–$6\frac{1}{2}$ lb per 280 lb sk of flour.

Straight dough system. The operations and typical time sequence in a 3 hr fermentation straight dough process are shown in the diagram (Fig. 32 p. 180).

The ingredients for a 1–sk dough would probably be 280 lb of flour, $3\frac{1}{2}$ lb of yeast, 5 lb of salt and 15–$15\frac{1}{2}$ gal of water. These are mixed, using water at a temperature that will bring the mixture to about 80°F. The yeast is dispersed in some of the water, and the salt dissolved in another portion; the yeast suspension,

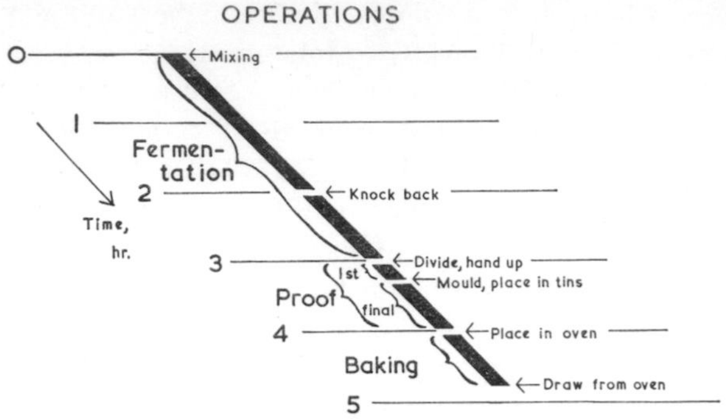

OPERATIONS

Fig. 32. Operations and typical time sequence in a 3 hr fermentation straight dough procedure.

the salt solution, and the rest of the water are then blended with the flour, and the dough set aside while fermentation proceeds.

After about 2 hr the dough is "knocked back", that is, manipulated to push out the gas that has been evolved. The process of knocking back may even out the temperature and give more thorough mixing in the case of bulk doughs.

After another hour's rising, the dough is divided into loaf-sized portions and these are roughly shaped. The doughs rest for 10–15 min ("1st proof") and are then moulded into the final shape, i.e. mechanically worked to tighten the dough so that the gas is better retained, and placed in tins. The doughs rest again, in the tins, for the final proof of 40–45 min and are then baked in the oven, at a temperature of about 450–500°F, for about 45 min, with steam injected into the oven to produce a glaze on the crust.

The 3 hr fermentation process is the system mostly used in England. With fewer small bakeries operating, 10–12 hr overnight processes are less common now than formerly.

Sponge system. The straight dough system is used in England, but in the U.S.A., and to some extent in Scotland, a sponge system is used.

The sponge system differs from the straight dough system in

that only part of the flour is mixed at first with all the yeast and sufficient water to make a dough, which is allowed to ferment for some hours. The sponge (as this first dough is called) is then broken down, and the remainder of the flour, water and all the salt added to make a dough of the required consistency, which is given a short fermentation time only before proofing and baking. Details of procedure are to be found in *Breadmaking* by Bennion (1954).

The sponge system, being a longer process, requires less yeast than is used for a straight dough, and produces bread that has a fuller flavour. The disadvantage of the sponge system is that it is more complicated than the straight dough system.

No-time dough. This is a short system that is occasionally used in the U.S.A. and in Britain, particularly for emergency production. The dough is made, using a larger amount of yeast, e.g. 7 lb/sk, and a higher temperature, e.g. 86–89°F, than is usual for normal systems, and is immediately scaled off. Final moulding follows after about 15 min, and the doughs are proofed for 1 hr at 110°F. The bread has a coarse thick-walled crumb structure, and it stales rapidly.

Machine-baking. In small bakeries, all the processes of mixing, dividing, moulding, placing in the proofing cabinet and the oven and withdrawing from the same are carried out by hand. However, disadvantages of hand-processing are lack of uniformity in the products and excessive amount of labour involved. In large bakeries, machines carry out all these processes. Mixers are of the slow-speed open pan, or closed high-speed types; dough dividers divide the dough by volume; automatic provers have built-in controls giving correct temperature and relative humidity. In the smaller machine bakeries, peel or reel ovens have now largely replaced drawplate ovens. In larger bakeries, so-called "travelling ovens" are used. In these, the doughs are placed on endless bands which travel slowly through the oven, which is tunnel-shaped and possibly 100 ft or more in length.

Continuous doughmaking. A further stage in the mechanization of breadmaking is represented by the continuous doughmaking process, exemplified by the Wallace & Tiernan Do-Maker Process, based on the work of Baker, which is in use in the U.S.A. and elsewhere. In this process (see Anon., 1957), the flour, spouted from a hopper, is continuously mixed with a liquid or "brew" containing the yeast, salt, etc., in electronically regulated quantities. The dough is allowed no fermentation time, but instead is subjected to intense mechanical mixing whereby the correct conditions for baking are attained. In the absence of fermentation, it is essential to incorporate a fair quantity of oxidizing agent in the dough. The dough is extruded through a pipe, cut off into loaf-sized portions, proofed, and baked. The Do-Maker Process gives bread with a characteristic and very even crumb texture. Considerable time is saved, in comparison with normal processes.

Chorleywood Bread Process. The Chorleywood Bread Process is a batch process of breadmaking, in which bulk fermentation of dough is replaced by the expenditure of intense mechanical energy during the mixing of the dough. It was developed by cereal scientists and bakers at the British Baking Industries Research Association, Chorleywood, Herts., England. The process is characterized by the expenditure of a considerable amount of work (0·4 h.p.-min/lb, or 5 W-hr/lb, or 40 joules/g) on the dough during a period of about 5 min; by chemical oxidation with ascorbic acid at a high level, viz. 75 ppm; by the addition of fat (about 0·7%) and extra water (3·5% more than normal, based on flour weight) together with the absence of any pre-ferment or liquid ferment. The quantity of yeast used is somewhat higher than for normal processes. It is said that bread made by the process is indistinguishable in flavour or crumb structure from bread made by normal processes, and that it stales less rapidly (Chamberlain *et al.*, 1962; Axford *et al.*, 1963).

Advantages claimed for the system, besides the avoidance of bulk fermentation, are an additional yield of about 7 lb of dough per 100 lb of flour, a saving in time and space, and greater amenability to control. A very large proportion of the bread baked in Britain is now made by this process.

Wholemeal bread. A short fermentation system is generally used for wholemeal bread, because the quantity of enzymes in wholemeal is larger than in white flour. The dough, made at 74°F, is allowed to ferment for 1 hr before knocking back, plus 30 min to scaling and moulding.

Rope. Freshly milled flour contains bacteria and mould spores, but these normally cause no trouble in bread under ordinary conditions of baking and storage. The vegetative forms of the bacteria are killed at oven temperature, but spores of some of the bacteria survive, and may proliferate in the loaf if conditions are favourable, causing a disease of the bread known as "rope". Ropy bread is characterized by the presence in the crumb of yellow-brown sticky spots and an objectionable adour. The organisms responsible are members of the *B. subtilis-mesentericus* group. Proliferation of the bacteria is discouraged by acid conditions in the dough (e.g. by addition of $1\frac{1}{2}$–2 lb of ACP or 2 pints of 12% acetic acid per 280 lb sk of flour) and by rapid cooling of the bread.

The Preservatives in Food Regulations 1962 permit (in Britain) the addition to bread of propionic acid and some of its salts to prevent development of mould in bread—a possibility in hot weather.

Bread cooling. The cooling of bread is a problem in mechanized production, particularly when the bread is to be sliced and/or wrapped before sale. Bread leaves the oven with the crumb at a temperature slightly below 212°F, and with about 45% m.c. at the centre. The crust is hotter but much drier (1–2% m.c.) and cools rapidly. During cooling, moisture moves from the interior outwards towards the crust and thence to the atmosphere. If the moisture content of the crust rises considerably during cooling, the texture of the crust becomes leathery and tough, and the attractive crispness of freshly baked bread is lost.

Extensive drying during cooling results in weight loss (and possible contravention of the Weights and Measures Act) and in poor crumb characteristics. The aim in cooling is therefore to lower the temperature without much loss of moisture. This may be achieved by subjecting the loaves to a counter-current of air conditioned to about 70°F and 80% r.h. The time taken for cooling 1¾-lb loaves by this method is 2–3 hr. In another method, using a two-stage vacuum cooling, the loaves can be cooled in 35 min. In the first stage, the bread is cooled to about 135°F by a counter-current of conditioned air during 30 min; this is followed by a vacuum section in which the temperature falls rapidly—in 3 or 4 min.

Bread staling. Staling of bread crumb is not a drying-out process; loss of moisture is not involved in true crumb staling. Workers, from Katz (1928) onwards, have agreed that the basic cause of staling is the transformation of starch from one chemical form to another. Katz claimed that starch slowly changed at temperatures below 55°C (131°F) from an alpha to a beta form, the latter binding considerably less water than the former, and that this change led to a rapid hardening, and to shrinkage of the starch granules away from the gluten skeleton with which they are associated, with consequent development of crumbliness. Katz found that staling can be prevented if bread is stored at temperatures above 55°C (although this leads to loss of crispness and the possibility of rope development) or at −20°C.

Schoch and French (1945) concluded that bread staling was due to internal coacervation, i.e. heat-reversible aggregation of the amylopectin (branched-chain) portion of the starch, and was not concerned with any change in the amylose (straight-chain) portion of the starch, as the latter is insolubilized by irreversible retrogradation during baking and hence cannot influence the staling that occurs subsequently. Reviews of the subject have been published by Hutchinson and Fisher (1937), Geddes and Bice (1946), Hintzer (1950), and Radley (1953). See also Hlynka (1964).

Biscuit baking. In the manufacture of biscuits, flour, water, fat, sugar and other ingredients are mixed together. The dough is rested for a time and then passed between rollers to make a sheet, from which the biscuit shapes are stamped out. For making biscuits, it is important that the flour used should produce dough that has particular rheological properties, viz. low spring and high extensibility (cf. p. 149).

REFERENCES

ANON. (1957), Bread making process now available, *Northw. Miller* **257** (13): 13.

AXFORD, D. W. E., CHAMBERLAIN, N., COLLINS, T. H. and ELTON, G. A. H. (1963), The Chorleywood process, *Cereal Sci. Today* **8**: 265.

BENNION, E. B. (1954), *Breadmaking*, 3rd edition, Oxford University Press.

BERTRAM, G. L. (1953), Studies on crust colour, *Cereal Chem.* **30**: 127.

CHAMBERLAIN, N., COLLINS, T. H. and ELTON, G. A. H. (1962), The Chorleywood Bread Process, *Bakers' Dig.* **36** (5): 52.

GEDDES, W. F. and BICE, C. W. (1946), *The Role of Starch in Bread Staling*, Quartermaster Food and Container Institute, U.S.A.

HINTZER, H. M. R. (1950), *The Staling of Bread*, Nordish Cerealkjemiker-forenings, Kongress, Bergen.

HLYNKA, I. (1964) (Ed.), *Wheat: Chemistry and Technology*, Amer. Assoc. Cereal Chem., St. Paul, Minn., U.S.A.

HUTCHINSON, J. B. and FISHER, E. A. (1937), The staling and keeping quality of bread, *Bakers Nat. Ass. Rev.* **54**: 563.

JONES, C. R. (1940), The production of mechanically damaged starch in milling as a governing factor in the diastatic activity of flour, *Cereal Chem.* **17**: 133.

KATZ, J. R. (1928), Gelatinization and retrogradation of starch in relation to the problem of bread staling, in WALTON, R. P. (1928), *A Comprehensive Survey of Starch Chemistry* **1**: 100, Chemical Catalog Co. Inc., New York.

MINISTRY OF AGRICULTURE, FISHERIES AND FOOD (1962), *Preservatives in Food Regulations 1962*, Statutory Instruments 1962, No. 1532, H.M.S.O., London.

PYLER, E. J. (1952), *Baking Science and Technology*, Siebel Publ. Co. Chicago.

RADLEY, J. A. (1953), *Starch and its Derivatives*, Chapman & Hall, London.

SCHOCH, T. J. and FRENCH, D. (1945), *Fundamental Studies on Starch Retrogradation*, Office of the Quartermaster General, U.S.A.

FURTHER READING

BENNETT, R., *Principles of Baking*, Cambridge Food Science Course, 1951.

CORRESPONDENCE COURSE, *Milling*, 8 Feb. 1963.

HALTON, P., Wheat for biscuit flours, *Food Manuf.* **27**: 149, 1952.

HALTON, P., The development of dough by mechanical action and oxidation, *Milling* **138**: 66, 1962.

JONES, C. R., GREER, E. N., THOMLINSON, J. and BAKER, G. J., Technology of the production of increased starch damage in flour milling, *Milling* **137**: 58, 80, 1961.

KENT-JONES, D. W. and MITCHELL, E. F., *Practice and Science of Breadmaking*, 3rd edition, Northern Publ. Co, Ltd., Liverpool, 1962.

SCIENCE EDITOR, Bread cooling, *Milling* **140**: 106, 1963.
SCIENCE EDITOR, Yeast growth in bread doughs, *Milling* **139**: 104, 1962.
SPEIGHT, J., Bread through the ages, *Milling* **139**: 160, 1962.
TRUM, G. W. and SNYDER, E. G., Continuous dough processing, *Cereal Sci. Today* **7**: 344, 1962.

CHAPTER II

BREAKFAST CEREALS AND OTHER PROCESSED PRODUCTS

BREAKFAST CEREALS

Breakfast cereal foods can be classified according to (a) the amount of domestic cooking required, (b) the form of the product or dish, (c) the cereal used as raw material. Types that require no cooking are called ready-to-eat cereals.

Cooking of cereals

All cereals contain a large proportion of starch. In its natural form, the starch is insoluble, tasteless, and unsuited for human consumption. To make it digestible and acceptable it must be cooked. In the case of ready-to-eat cereals, the cooking is carried out during manufacture.

If the cereal is cooked with excess of water and only moderate heat, as in *boiling*, the starch gelatinizes and becomes susceptible to starch-dissolving enzymes of the digestive system. If cooked with a minimum of water, or without water, but at higher temperature, as in *toasting*, non-enzymic browning reaction between protein and reducing carbohydrate may occur, and there may be some dextrinization of the starch.

Porridge. The manufacture of oatmeal and porridge oats (rolled oats, oatflakes) will be described in Ch. 13. Oatmeal is milled by a process that includes no cooking (unless the oats are stabilized: cf. **p.** 216), and the starch in oatmeal is ungelatinized; moreover, the particles of oatmeal are relatively coarse in size. Consequently, porridge made from coarse oatmeal requires prolonged domestic cooking to bring about gelatinization of the starch. Oatmeal of flour fineness cooks quickly, but the cooked product is devoid of the granular structure associated with the best Scotch porridge.

Rolled oats are partially cooked during manufacture; the pinhead meal from which rolled oats is made is softened by treatment with steam, and, in this plastic condition, is flattened on flaking rolls. About one-third of the starch becomes gelatinized. Thus, porridge made from rolled oats requires only a brief cooking time to complete the process of starch gelatinization.

Ready-cooked porridge. A product from which a type of porridge can be made merely by stirring with hot or boiling water in the bowl, called "Porridge without the pot", consists of oat flakes of special type. As compared with ordinary rolled oats, these flakes are thinner, stronger, and contain starch which is more completely gelatinized. They could be manufactured by steaming the pinhead meal at a somewhat higher moisture content than normal, rolling at greater roll pressures than normal, and using heated flaking rolls.

Another type of porridge mix, known as Readybrek, consists of a blend of two types of flakes in approximately equal proportions: (a) ordinary rolled oats made from small pinhead meal, and (b) very thin flakes of a roller-dried batter of oat flour and water, similar to products of this nature used for infant feeding. When this porridge mix is stirred with hot water, the thin flakes form a smooth paste while the rolled oats, which do not completely disperse, provide a chewy constituent and give body to the porridge.

Ready-to-eat cereals

These comprise flaked, puffed, shredded and granular products, generally made from wheat, maize or rice, although oats and barley are also used. The basic cereal may be enriched with sugar, syrup, honey or malt extract. All types are prepared by processes which tend to cause dextrinization rather than gelatinization of the starch.

Flaked products. Wheat, maize (for "corn flakes") or rice are the cereals generally used. Whole wheat or rice grain is cleaned, conditioned to a suitable moisture content, and lightly rolled between smooth rolls to fracture the outer layers. Moisture and flavourings can then penetrate the grain more readily, and this helps in the cooking. Whole wheat or rice, so prepared, or maize grits, is cooked, often at elevated pressure, and the flavourings (malt, salt, sugar, etc.) added. The cooked cereal is dried to 15–20% m.c. and rested for 24–72 hr while conditioning takes place. The conditioned grain is flaked on heavy flaking rolls, toasted in a tunnel or "travelling" oven, cooled and packeted.

Puffed product. Whole grain wheat, rice or oats is prepared by cleaning, conditioning and depericarping (by a wet scouring process). Alternatively, dough made from maize meal or oat flour blended with tapioca or rye flour is made to a stiff consistency (30–35% m.c.), with the addition of sugar, salt and sometimes oil. It is cooked for 20 min at 20 lb/in^2 pressure, dried to 14–16% m.c., and pelleted by extrusion through a die. A batch of the prepared grain or pelleted dough is fed into a pressure chamber, which is then sealed, and heated externally and by injection of steam so that the internal pressure rapidly builds up to about 200 lb/in^2 and is then suddenly released by opening the chamber (called a "Puffing gun"). Expansion of water vapour on release of the pressure blows up the grains or pellets to several times their original size. The puffed product is dried to 3% m.c. by toasting, then is cooled and packaged.

For satisfactory puffing, the material at the moment before expansion requires cohesion to prevent shattering and elasticity to permit expansion. The balance between these two characteristics can be varied by adding starch, which has cohesive properties.

Shredded product. A white starchy wheat is used for shredding. The whole grain is cleaned and then cooked with water by application of external heat and injection of steam. The cooking conditions are such that the cooked grain is soft and rubbery, the moisture content is about 43%, and the starch is fully gelatinized. The cooked grain is cooled, and rested for some hours to condition. The conditioned grain is fed to shredders consisting of a pair of metal rolls: one is smooth, the other has circular grooves between which the material emerges as long parallel shreds. The shreds fall onto a slowly travelling band, and a thick mat is built up by superimposition of several layers. The mat is cut into tablets and the latter baked for 20 min at 500°F in a gas-heated revolving oven. After baking, the tablets are dried to 1% m.c., cooled, passed through a metal detector, and packaged.

Granular product. A yeasted dough is made from a fine wholemeal or long extraction wheaten flour and malted barley flour, with added salt. The dough is fermented for about 5 hr and from it large loaves are baked. These are broken up, dried and ground to a standard degree of fineness.

Sugar-coated cereals. Flaked or puffed cereals, prepared as described, are sometimes coated with sugar or candy. The process described in Brit. Pat. Spec. No. 754,771 uses a sucrose syrup containing 1–8% of other sugars (e.g. honey) to provide a hard transparent coating that does not become sticky even under humid conditions. The sugar content of cornflakes was raised from 9 to 33% by the coating process, that of puffed wheat from 2 to 53%.

Keeping quality of breakfast cereals

The keeping quality of the prepared product depends to a large extent on the content and keeping quality of the fat contained in it. Thus, products made from cereals having a low oil content (wheat, barley, rice, maize grits: oil content 1·5–2·0%) have an advantage in keeping quality over products made from oats (oil content: 4–11%, average 7%). Whole maize has high oil content (4·4%), but most of the oil is contained in the germ which is removed in making grits (cf. p. 247).

The keeping quality of the fat depends on its degree of unsaturation, the presence or absence of antioxidants and prooxidants, the time and temperature of the treatment, the moisture content of the material when treated, and the conditions of storage. Severe heat treatment, as in toasting or puffing, may destroy antioxidants or induce formation of pro-oxidants, stability of the fat being progressively reduced as treatment temperature is raised, treatment time lengthened, or moisture content of the material at the time of treatment lowered. On the other hand, momentary high temperature treatment, as at the surface of a hot roll in the roller-drying of a batter, may produce new antioxidants by interaction of protein and sugar (non-enzymic browning, or Maillard reaction): such a reaction is known to occur, for example, in the roller-drying of milk, and may be the explanation of the improved antioxidant activity of oat products made from oats after steam treatment for lipase in-activation (cf. pp. 216 and 223).

The addition of synthetic antioxidants to prepared breakfast cereals is not at present permitted in Britain (cf. p. 195).

Nutritive value of breakfast cereals

The chemical composition of some ready-to-eat breakfast cereals is shown in Table 45.

Shredded wheat, made from low-protein, soft wheat, has a protein content considerably lower than that of puffed wheat, which is made from a high-protein, hard wheat, such as Manitoba.

All cereal products are deficient in the amino acid lysine, but the deficiency may be relatively greater in ready-to-eat cereals than in bread because of the changes that occur in the protein at high temperature of treatment. However, lysine deficiency is of less importance in ready-to-eat breakfast cereals than in bread because the former are generally consumed with milk, which is a good source of lysine. Moreover, some ready-to-eat breakfast cereals have a protein supplementation.

The calorie value of ready-to-eat cereals is higher than that of bread (243 cal/100 g), largely on account of the relatively lower moisture content of the former. Compared at equal moisture contents, the difference in calorie value is small.

The processes involved in the manufacture of ready-to-eat

TABLE 45

CHEMICAL COMPOSITION OF READY-TO-EAT BREAKFAST FOODS (PER 100 G)

Food	Water (g)	Sugar (g)	Available Carbohydrates (g)	Protein N×5.7 (g)	Protein N×6.25 (g)	Fat (g)	Ash (g)	Fibre (g)	Calories	Source of data
Grapenuts	4.8	11.9	75.2	11.7	—	3.0	2.6	1.7	358	1
Shredded wheat	8.0*	Trace	79.0	9.7	—	2.8	1.6	1.8	362	
Weetabix	8.2*	5.9	77.0	10.9	—	1.9	—	—	351	
Kellogg's:										
All-Bran	2.3	17.2	67.0	13.2	—	3.3	7.7	6.6	350	2
Whole Wheat Flakes	3.0	17.1	81.2	8.4	—	1.7	4.1	1.7	374	
Cornflakes	2.0	9.4	88.0	—	7.5	0.5	2.0	0.5	367	
Rice Krispies	2.0	11.5	87.0	—	6.0	0.5	2.0	0.4	355	
Frosties	3.0	43.1	90.3	—	4.5	0.2	2.0	0.3	365	
Sugar Smacks	3.0	54.5	89.0	5.0	—	1.5	2.0	1.0	375	
Sugar Ricicles	2.4	39.8	90.7	—	3.7	0.6	2.2	0.4	383	
Cocopops	3.0	34.7	84.4	—	5.4	4.8	2.0	0.4	390	
Special K	3.0	10.5	73.0	—	20.0†	0.5	2.0	0.5	360	
Quaker Oats:										
Puffed Wheat	4.0	2.2	75.2	14.1	—	2.0	1.5	2.0	348	3
Oat Krunchies	3.5	15.0	75.0	10.1	—	5.6	4.5	0.6	385	
Sugar Puffs	3.2	51.0	86.5	6.8	—	1.0	0.8	1.0	371	

* m.c. when analysed. The m.c. when freshly packed may be 2–3%.
† Enriched to this level.

Sources of data: 1. Data for ash and fibre from M. B. Jacobs (1944). Other data from McCance and Widdowson (1960). (Fat by von Lieberman's method.) 2. Data by courtesy of the Kellogg Company of Great Britain Ltd. (1964). 3. Data by courtesy of Quaker Oats Ltd. (1964).

cereals cause partial hydrolysis of phytic acid (cf. p. 163); the amount of destruction increases at high pressures: about 70% is destroyed in puffing, about 33% in flaking.

The contents of some of the B vitamins in ready-to-eat breakfast foods is shown in Table 46. Information concerning other vitamins present is meagre. About 50% of the vitamin B_1 is destroyed during the manufacture of shredded wheat, and nearly 100% is destroyed during puffing and flaking. These processes have little effect on riboflavin and niacin. Certain ready-to-eat cereals manufactured in Britain are enriched with vitamins, as shown in Table 46. Weetabix, in addition, is enriched with iron (to 60 $\mu g/g$ total content).

TABLE 46

VITAMIN CONTENT OF BREAKFAST FOODS

($\mu g/g$)

Food	Vitamin B_1	Riboflavin	Niacin	Source of data
Puffed barley	0	1·1	76	1
Shredded wheat	2·6	1·5	49	
Kellogg's:				
All-Bran	10·0	14·3	107	
Whole Wheat Flakes	7·0	13·4	103	
Cornflakes	6·0*	10·7*	71*	
Rice Krispies	6·0*	10·7*	71*	
Frosties	3·6	6·4	43	2
Sugar Smacks	Trace	0·5	15	
Sugar Ricicles	3·6	6·4	43	
Cocopops	3·6	6·4	43	
Special K	6·0*	10·7*	71*	
Quaker Oats:				
Puffed Wheat	12·5*	1·2–1·4	46–55	3
Oat Krunchies	2·8			

* Enriched to these levels.

Sources of data: 1. Booth *et al.* (1945). 2. Courtesy of the Kellogg Company of Great Britain Ltd. (1964). 3. Data for vitamin B_1 by courtesy of Quaker Oats Ltd. (1964); other data from Booth *et al.* (1945).

MACARONI

Macaroni and other pasta products (spaghetti, vermicelli, noodles) are made from semolina (cf. p. 116) which is milled from hard wheat by a special process. The highest quality pasta products are made from durum wheat alone; other hard wheat, e.g. Manitoba or HRS, can be substituted, but only at the expense of quality.

In order to obtain the maximum yield of clean semolina and the minimum amount of flour (which is a by-product in this process), a system with 6 or 7 breaks is used, grinding quite lightly in the early stages. The semolina released is closely graded into a large number of fractions, each with a narrow particle size range, and is purified by an elaborate system in order to free the semolina from bran fragments as completely as possible. Using such a system, a good durum wheat would yield 65% of semolina and only 10% of flour.

The qualities looked for in semolina for macaroni manufacture are brightness, i.e. absence of bran specks, and yellow pigmentation. As the lipoxidase enzyme in semolina may destroy the yellow pigment during subsequent doughmaking, a low lipoxidase activity in the wheat is desirable.

In the manufacture of macaroni, the semolina is made into a stiff dough, using $2\frac{1}{2}$–3 gal of water at 90–100°F per 100 lb of semolina, and mixing for 10–15 min. After resting, the dough is extruded at heavy pressure (e.g. 3000 lb/in^2) through the die of a press to make the tube- or strap-shaped products. Heavy pressure is necessary to ensure that the product is translucent: small air bubbles are squeezed out by the heavy pressure.

Finally, the product is dried. In Italy, drying may be done outdoors, but elsewhere special drying cabinets are used in which the temperature and relative humidity of the air can be carefully controlled. The rate of drying must be correct: drying at too low a rate may lead to development of moulds, discoloration and souring, whilst drying too rapidly may cause cracking ("checking" in the U.S.A.) and curling.

A method of drying recommended by Kent-Jones and Amos (1957, p. 380) is to allow drying at about 90°F for the first 15–20 hr in order to induce the formation of a slight crust, which discourages mould growth, and then to complete the drying at a lower temperature and at about 75% r.h. Total drying time is generally 30–80 hr.

Good-quality macaroni should be creamy in colour, free from cracks, and somewhat flexible; when broken, the fracture should appear glassy. When boiled in water for 10 min, the macaroni should swell to twice its original volume, but should retain its shape and firmness without becoming pasty and without disintegrating.

The *per caput* consumption of pasta products in 1960 was (in kg) 30 in Italy, 12 in Greece, 5–9 in Switzerland, Portugal, France, 2–3 in most other European countries, Canada, the U.S.A., 0·4 in the U.K.

CAKE PREMIXES

Cake premixes sold in Britain normally contain, in powder form, all the ingredients required for a cake, viz. flour, fat, sugar, baking powder, milk powder, eggs, flavouring and colour, and need only the addition of water before baking. However, some cake premixes, particularly those sold in the U.S.A., omit the eggs and/or the milk, because lighter cakes of larger volume can be made by the use of fresh eggs and milk instead of the dried ingredients.

In preparing the premixes, the dry ingredients are measured out by automatic measurers and conveyed, often pneumatically, to a mixing bin, mixed, and then entoleted (cf. p. 141) to ensure freedom from insect infestation. The fat is then added, and the mixture packaged. If fruit is included in the formula it is generally contained in a separate cellophane-wrapped package enclosed in the carton.

The type of flour used must be suitable for the particular product, flours of the high ratio type (cf. p. 150) generally being used. The fat must have correct plasticity and adequate stability to resist oxidation. The addition of certain antioxidants to fat, to improve stability, is allowed in Britain, the U.S.A., and elsewhere. Those at present allowed in Britain for addition to fats other than butter are propyl gallate up to 0·01% or butylated hydroxy anisole (BHA) up to 0·02% (calculated on the fat). Those allowed in the U.S.A. are resin guaiac (0·1%), tocopherols (0·03%), lecithin, citric acid (0·01%), nordihydroguaiaretic acid (0·01%), pyrogallate (0·01%), and BHA (0·2%).

RICE SUBSTITUTES

Bulgur

Bulgur consists of parboiled whole or crushed wheat grains, and is used as a substitute for rice, e.g. in pilaf, an eastern European dish consisting of wheat, meat, oil and herbs cooked together. A product resembling the wheat portion of this dish was developed in the U.S.A. in 1945 as an outlet for part of the U.S. wheat surplus.

In one method for the manufacture of bulgur, described by Schäfer (1962), cleaned white or red soft wheat, preferably decorticated, is cooked by a multistage process in which the moisture content is gradually increased by spraying with water and the temperature gradually raised. Eventually, when the moisture content is 40%, the wheat is heated at 94°C and then steamed for $1\frac{1}{2}$ min at a controlled pressure so that the cooked product is gummy and starchy, but not discoloured by overheating. The moisture content is then reduced to about 10% by drying with air at 66°C, and the dried cooked wheat is pearled or cracked. One brand of the whole grain product is called Rediwheat in the U.S.A.; the crushed is Cracked-Bulgur.

Bulgur is sent from the U.S.A. to peoples in the Far East as part of the programme of American aid to famine areas. The staple food of people in these areas had always been boiled rice; the process of breadmaking was unknown; wheat and wheat flour were therefore unacceptable foods. Bulgur provided a cheap food that was acceptable because it could be cooked in the same way as rice and superficially resembled it.

Bulgur has good storage properties and resists attack by insects and mites.

Ricena

Ricena is a rice substitute, originating in Australia, made from wheat by a relatively inexpensive process which gives a yield of about 65%. The product sells at the price of low grade rice, although the protein, iron and vitamin B_1 contents of ricena exceed those of milled rice.

REFERENCES

BOOTH, R. G., MORAN, T. and PRINGLE, W. J. S. (1945), The nutritive value of ready-to-eat breakfast cereals, *J. Soc. Chem. Ind.* **64**: 302.

GENERAL FOODS CORPORATION INC. (1954), Coated cereals, *Brit. Pat. Spec.* 754,771.

JACOBS, M. B. (1944), *Chemistry and Technology of Food and Food Products*, Interscience Publ., New York.

KENT-JONES, D. W. and AMOS, A. J. (1957), *Modern Cereal Chemistry*, 5th edition, Northern Publ. Co. Ltd., Liverpool.

McCANCE, R. A. and WIDDOWSON, E. M. (1960), *The Composition of Foods*, Med. Res. Coun., Spec. Rep. Ser. 297, H.M.S.O., London.

MINISTRY OF AGRICULTURE, FISHERIES AND FOOD (1958), *The Antioxidants in Food Regulations 1958*, Statutory Instruments 1958, No. 1454, H.M.S.O., London.

SCHÄFER, W. (1962), Bulgur for underdeveloped countries, *Milling* **139**: 688.

FURTHER READING

BRIT. PAT. SPEC. Nos. 618,524; 677,088; 719,870; 735,530; 759,478; 919,906.

MARTIN, H. F., Factors in the development of oxidative rancidity in ready-to-eat crisp oatflakes, *J. Sci. Fd. Agric.* **9**: 817, 1958.

RADLEY, J. A., The chemistry and physics of macaroni products, *Food Manuf.* **27**: 327, 369, 406, 436, 481, 1952; **28**: 11, 1953.

SCIENCE EDITOR, Prepared breakfast cereals, *Milling* **139**: 584, 1962.

SPEIGHT, J., Pasta alimentare, *Milling* **139**: 110, 1962.

CHAPTER 12

BARLEY:
PROCESSING, NUTRITIONAL
ATTRIBUTES, TECHNOLOGICAL USES

HISTORICAL

The cultivation of barley (*Hordeum* sp.) reaches far back into human history; it was known to the ancient Egyptians, and grains of 6-row barley have been discovered in Egypt dating from pre-dynastic and early dynastic periods. Barley is mentioned in the Book of Exodus in connection with the ten plagues.

Barley was used as a bread grain by the ancient Greeks and Romans. Hunter (1928) illustrates Greek coins dating from 413 to 50 B.C. which incorporate ears or grains of barley into their design. Barley was the general food of the Roman gladiators, who were known as the *hordearii* (Percival, 1921). Calcined remains of cakes made from coarsely ground grain of barley and *Triticum monococcum*, dating from the Stone Age, have been found in Switzerland.

Bread made from barley and rye flour formed the staple diet of the country peasants and the poorer people in England in the fifteenth century (cf. p. 228), while nobles ate wheaten bread. As wheat and oats became more generally available, and with the cultivation of potatoes, barley ceased to be used for bread-making.

USES

The principal uses for barley are as feed for animals, particularly pigs, in the form of barley meal, for malting and brewing in the manufacture of beer, and for distilling in whisky manufacture (cf. p. 204). Barley finds little use for human food in Europe and North America, but is widely used for this purpose in Asian countries. Even there, however, its use for human food is declining as preferred grains become more plentiful.

Utilization. Domestic utilization of barley in 1961–2 in certain countries is shown in Table 47.

TABLE 47

DOMESTIC UTILIZATION OF BARLEY IN 1961–2*

Country	Percentage of total consumption			
	Human food	Malting, milling, distilling	Animal feed	Seed
United States	25		68	7
Canada	12		81	7
Australia	46		40	14
United Kingdom	22		73	5
France	0·1	7	86	7
Japan (1960/1)	63	10	25	2

* Sources *Grain Bulletin*; *Grain Crops* (see footnote of p. 1).

MALTING AND BREWING

For malting, both 2-row and 6-row hulled types (cf. p. 9) are suitable: the former is generally used in Europe, the latter in North America. A low nitrogen content is required in malting barley and this limits the use of nitrogenous fertilizers. Distinct varieties were formerly grown for malting and for feeding, but the recent development of higher-yielding varieties that are suitable for malting has provided a surplus of these types for other purposes, viz. milling for human food, distilling, and animal feed.

The manufacture of beer from barley comprises two major processes: malting and brewing. Malting is a controlled germination process which is concerned with the "modification" of the grains, i.e. the liberation of starch granules from the endosperm cell matrix by enzymes which become active during germination. Brewing is the process of converting the starch to an alcoholic solution, first by transforming the starch to sugar, then by fermenting the sugar to alcohol by means of yeast. About 75% of the starch is generally fermented to alcohol.

Malting

The sequence of operations in malting is as follows:
1. Kiln drying.
2. Screening (cleaning the grain).
3. Storage.
4. Steeping.
5. Draining.
6. Spreading on the malting floor.
7. Turning or ploughing.
8. Drying in malt kiln.
9. Screening (removal of malt culms).

The grain is first dried in a kiln or drum dryer to between 10 and 14% moisture content. At this level of moisture content the grain can be safely stored. The kilning process sometimes improves malting quality by accelerating maturation of the grain. The drying also checks fungal and bacterial activity, reduces the rate of respiration of the grain, and hence reduces the respiratory loss.

The cleaning process embodies machinery operating on principles similar to those used in wheat cleaning (Ch. 6). Schuster (1962) recommends that cleaning should precede drying.

After drying and cleaning, the grain is stored in bulk for at least three weeks before malting. Freshly harvested barley exhibits a degree of dormancy and cannot be malted immediately (cf. p. 87). The period of storage allows the secondary ripening process to occur, so that the grain can attain its full germinative power.

The malting process proper begins when the barley grain is steeped in water. Time required for steeping depends on temperature and degree of aeration of the steep water. A temperature of 10–12°C is recommended, with steeping times of 40–60 hr in

England, 50–80 hr on the Continent (Schuster, 1962). The object of malting is to provide the conditions in which natural germination can proceed until the optimum enzymic activity has been developed. The process is designed to minimize the loss which is inevitable when grain germinates.

The grain germinates, if it is non-dormant, when water, in sufficient quantity to raise the moisture content to 42–46%, and oxygen are absorbed.

After steeping, the surplus water is drained off from the grain which is then spread on the malting floor in heaps or "couches" for a period of 8–12 days, while germination takes place. During this time the plumule grows to $\frac{1}{2}$–$\frac{2}{3}$ the length of the grain, an extensive root system develops, and modification of the endosperm proceeds.

Various enzymes are liberated during germination: among the first of these is cytase, an enzyme that dissolves the material binding the endosperm cell walls, and helps to liberate the starch granules contained in the endosperm cells. Other enzymes that become active in the early stages include phosphatase, phytase, hemicellulase, and protease. Amylases become active at a later stage.

Accompanying the increased enzymic activity which mobilizes the high molecular weight substances in the seed, and permits their translocation to the embryo to make new tissues in growth, there is considerable increase in the rate of respiration of the grain—the process in which starchy material is converted (via sugar) to carbon dioxide and water. The respiratory loss of dry matter during malting is generally 4–8%, depending on the length of time the grain remains on the malting floor. The loss is minimized when germination is rapid and uniform. Starch is the most valuable portion of the grain, and a long time on the malting floor involves greater starch loss.

The degree of modification required depends on the type of beer brewed. Less modification is required in grain for pale ales than for dark ales.

When the grain has been modified sufficiently it is dried in a malt kiln, first at a low temperature and later at a temperature high enough to suspend enzymic activity without destroying the enzymes. Finally, the dried grain is screened to remove the rootlets—called malt culms—which are now dry and brittle. The screened product is malt.

Malting condition. The condition of barley for malting is a matter of great importance, having probably a greater effect on the yield and quality of the products than the condition of wheat has on the flour milled from it.

The characteristics of good malting barley, as defined by Wrightson and Patrick Wright (1908), are as follows:

1. High germination capacity, not less than 90%.
2. High starch and low nitrogen contents.
3. Thin, reticulated skin (husk).
4. Mealy, floury appearance (i.e. not steely) when cut across.
5. Uniform pale yellow colour.
6. Freedom from broken grains.
7. Absence of musty odour.

The need for high germination capacity follows from the foregoing description of the malting process. Loss of germination capacity could result from careless threshing, from drying at too high a temperature (cf. p. 90), or from heating caused by storage at too high a moisture content (cf. p. 89). The uniform pale yellow colour and the absence of musty odour give some assurance of satisfactory drying and storage.

Uniform, complete germination requires the grain to be ripe; unripe barley has a smooth unwrinkled skin, whereas the skin of the fully ripe grain is finely reticulated. The husk of the grains of malting barley should be intact because damage to the husks allows moulds to develop during malting. The husk in broken grains is, of course, not intact.

If the barley possesses these desirable characteristics, then the yield of malt extract that may be expected is related directly to the starch content and the grain size, and inversely to the protein content.

High-nitrogen barley is unsuitable for malting because yield of extract is reduced and its quality impaired:

1. High nitrogen content necessarily implies low starch content. It is the starch—converted to maltose and dextrins—that is the most important fermentable constituent of the malt extract.

2. High-nitrogen barley requires a longer time for modification, and a longer malting time entails more rootlet development and greater respiratory and metabolic loss.

3. High-nitrogen barley does not modify to the same extent as low-nitrogen barley, however long a malting time is allowed.

4. The malt from high-nitrogen barley contains relatively more soluble protein or albuminoid material than that from low-nitrogen barley; this soluble protein will pass into the extract and may impair the keeping quality or "stability" of the beer. Furthermore, development of bacteria is more likely in liquor with a high albuminoid content.

High nitrogen content in barley is associated with a steely or flinty appearance of the endosperm when the grain is cut across (cf. p. 65); the endosperm of low-nitrogen barley has a mealy or floury appearance.

The average nitrogen content of barley is about $1 \cdot 5\%$; some 38% of this appears in the beer in the form of soluble nitrogen compounds, the proportion of the total N entering the beer being somewhat larger from 2-row than from 6-row.

Brewing

This is the second main process in the manufacture of beer, involving the fermentation of the malt extract.

In British breweries the malt is ground and then steeped or mashed in hot water (150°F) in vessels called mash tuns, to gelatinize the starch and permit its conversion to fermentable sugar by the action of the malt diastase (amylase). During the steeping process, the husks of the barley grains sink to the bottom of the mash tun, and the sugar goes into solution in the steeping liquor. When steeping is completed, the extract, known as "wort", can be filtered off through the husks as a clear solution.

The filter bed of husks is then washed with hot water, a process called "sparging", to complete the extraction. Hops are added to the wort for flavouring purposes, and the mixture of wort and hops is boiled to sterilize it and extract the hops. The boiling also coagulates some of the nitrogenous material in the wort and this assists in clarifying the beer. The liquor is filtered through the spent hops and, when cold, is seeded with yeast and allowed to ferment for several days. The greater part of the yeast is then removed, enough being left to permit a secondary fermentation, again of several days, before final removal of the yeast.

Brewing adjuncts. Starchy materials that are cheaper than malt are sometimes used as brewing "adjuncts" to replace a proportion of the malt. Materials so used include maize and wheat starches, barley and maize flakes, and cooked maize and rice grits. Such adjuncts must not impede the fermentation process, nor have any undesirable effect on quality of the products; almost entire conversion of the starch to sugar is also desirable.

Wheat flour has been successfully used as a brewing adjunct. Birtwistle *et al.* (1962) found that flour of 7% protein content, milled from Cappelle wheat, used to the extent of 25% of a dry mix with brewing malt, did not impede filtration of the wort, gave conversion of flour starch to sugar equal to that obtained from malt, had an undetectable effect on flavour of the beer, and improved storage life and head retention of the beer and extraction of the hops by the wort.

LIQUORS DISTILLED FROM CEREALS

Whisky is obtained from the distillation of a fermented mash of cereal grain, and is aged in wooden barrels for two to eight years before sale.

Scotch pot-still or malt whisky is made only from malted barley; Scotch patent-still or grain whisky from barley and other unmalted, cereal grains. In the manufacturing process, beer, made as described above (except for the omission of hops), is distilled at about 140 proof. The distinctive smoky flavour of

Scotch whisky is due to the peat used for firing the malt kilns, and to the characteristics of the water used.

U.S. whisky is generally made from maize or rye. Straight rye whisky is made from a fermented mash containing a minimum of 51% of rye; straight bourbon whisky from at least 51% of maize. Irish whiskey is made from malted barley alone or with admixture of unmalted barley, wheat, rye or oats. Canadian whisky is produced from maize, wheat, rye or barley.

Gin is a distilled spirit flavoured with juniper or some other flavouring. Dutch gin is made by distilling a mash of which at least one-third is barley. The distillate is re-distilled with the flavourings. English and U.S. gins are made by re-distilling grain whisky with the flavourings.

MILLING OF BARLEY

Barley is milled to make blocked barley, pearl barley (Graupen, in German), barley groats (Grütze), barley flakes, and barley flour for human consumption. The hull or husk of barley is largely indigestible, and is not desirable in these products; its removal is an important part of the milling process.

Operations. The sequence of operations in barley milling may be summarized as follows:

Pearl barley:
1. Preliminary cleaning
2. Conditioning
3. Bleaching (not practised in Britain)
4. Blocking
5. Aspiration to remove husk
6. Size-grading by sifting
7. Cutting on groat cutter for barley groats
8. Pearling of blocked barley or large barley groats
9. Aspiration, grading, sifting
10. Polishing

Barley flakes:
11. Pre-damping of barley groats

12. Steam-cooking of barley groats or pearl barley
13. Flaking on flaking rolls
14. Drying flakes on hot-air drier

Barley flour:
15. Roller-milling of pearl or blocked barley

Barley is cleaned on machines similar to those used for wheat cleaning, viz. milling separators, trieur cylinders or indented discs, and aspirators (cf. Ch. 6). The sizes of sieve apertures and indents are modified for the larger size of barley grains in comparison with that of wheat.

Conditioning consists in adjustment of moisture content by drying or damping.

Bleaching. The bleaching of barley is not permitted by law in Britain, but is generally practised in Germany (cf. p. 208). Imported barley is preferred to domestic grain for milling in Germany, because of its greater hardness and yielding capacity, and it is the foreign barley, the endosperm of which has a bluish colour, which is said to require bleaching.

Blocked barley (or occasionally whole barley) is fed into a vertical fireclay or earthenware cylinder into which steam and sulphur dioxide gas are injected. The quantities used are 1–2% of moisture (from the steam) and about 0·04% of sulphur dioxide (equivalent to 0·02% of S). Sometimes a solution of sulphurous acid (H_2SO_3) or of sodium bisulphite ($Na_2S_2O_4$) is used instead of the sulphur dioxide gas. After this treatment, which takes 20–30 min, the barley is binned for 12–24 hr for the bleaching to take effect.

Blocking and pearling. Both blocking (shelling) and pearling (rounding) of barley are scouring or abrasive processes, differing from each other merely in degree of removal of the superficial layers of the grain. Blocking removes part of the husk: this process must be accomplished with the minimum of injury to the kernels; pearling removes the remainder of the husk and part of the endosperm.

Three types of blocking and pearling machine are in general

use: (1) a batch machine consisting of a large circular stone, faced with emery–cement composition, and rotating on a horizontal axis within a perforated metal cage; (2) a continuous-working machine of Swedish make consisting of a rotor faced with abrasive material rotating on a horizontal axis within a semi-circular stator lined with the same material, the distance between rotor and stator being adjustable; (3) a continuous-working machine comprising a pile of small circular stones rotating on a vertical axis within a metal sleeve, the annular space between the stones and the sleeve, occupied by the barley, being strongly aspirated.

The first and second types are in use in Britain; all three are in use on the Continent. The barley falls between the rotor and the stationary part of each machine; in bouncing from one surface to the other the husk is split or rubbed off. The degree of treatment, resulting in either blocking or pearling, is governed by the abrasiveness of the stone facing, by the distance between rotor and stator, and by the time of treatment in batch machines or the rate of throughput in continuous-working machines.

Aspiration of the blocked or pearled grain to remove the abraded portions, and cutting of the blocked barley into portions known as grits, closely resemble the corresponding processes to be described in more detail in connection with oatmeal milling (cf. p. 219). Cutting of the blocked barley is not commonly carried out in Britain, where it is the practice to pass the whole blocked barley grains to the pearling machine. In Germany, however, the blocked barley is first cut into grits, the grits graded by size, and then rounded in the pearling machine.

The pearl barley is polished on machines similar to those used for pearling, but equipped with stones made of hard white sandstone instead of emery composition. The average yield of pearl barley in Britain is 58% of the whole barley. Barley which is yellowish in colour after polishing is, in Germany, whitened by adding talc powder in the polishing.

Barley flakes are made from pearl barley by steaming and

flaking on large-diameter smooth rolls, as described below for oats (cf. p. 220).

There is an extensive pearl barley milling industry in Germany, with export trade to Britain, European countries and elsewhere. The system of milling used employs an elaborate scheme of grading and sizing into as many as 12 different sizes—a range unknown in Britain, where the usual grades are described as Pot Barley, First Pearl and Second Pearl. Barley products for export to Britain are made from unbleached barley, and the polishing with talc powder is omitted.

BARLEY IN THE BREADMAKING GRIST

During the Second World War, when supplies of imported wheat were restricted in Britain, the Government authorized the addition of variable quantities (up to 10% in 1943) of barley, or of barley and rye or oats, to the grist for making bread flour (cf. p. 165). For this purpose, the barley was generally blocked to remove most of the husk.

BARLEY FOR ANIMAL FEED

As barley husk contains some 34% of fibre, and is relatively indigestible, the preferred type of barley for animal feeding has a low husk content. Whereas a low protein content is favoured for malting, a high protein content is more desirable for feeding purposes.

Total digestible nutrients in barley are given as 77–79% by Geddes (1936); 86–89% for hulled, 89% for hull-less, by Hill (1933). Digestibility coefficients for constituents of ground barley are given by Morrison (1947) as 79% for protein, 80% for fat, 92% for carbohydrate, and 56% for fibre.

The nutritive value of barley as cattle feed is slightly lower than that of maize. Barley is, nevertheless, preferred to maize for certain animals, e.g. pigs, and it is also used extensively in compound feeds.

BARLEY FLOUR

Pearl barley is generally used in Britain for the manufacture of barley flour. The milling system uses fluted and smooth rolls and plansifters in much the same way as in flourmilling (cf. Ch. 7). Barley flour can also be milled from blocked barley or from whole barley, but in the latter case due allowance must be made for the greatly increased quantity of by-products which would otherwise choke the system. Barley flour is also a by-product of the cutting, pearling and polishing processes.

An average extraction of 82% of barley flour is obtained from pearl barley representing 58% of the grain, i.e. an overall extraction rate of 48% based on the original grain. By using blocked barley, an overall extraction rate of 59% could be obtained, but the product would be considerably less pure.

NUTRITIVE VALUE

The chemical composition of milled barley products is shown in Table 48.

TABLE 48

CHEMICAL COMPOSITION OF MILLED BARLEY PRODUCTS

Material	Moisture (%)	Protein (%)	Fat (%)	Ash (%)	Fibre (%)	Carbohydrates (%)	Calories per 100 g	Source of data*
Pearl barley	10·8	8·7	1·0	1·2	0·8	78·3	357	1
Barley flour	10·0	10·2	1·7	1·2	0·7	76·9	364	1
Barley husk	10·4	1·4	0·3	5·6	34·0	48·3	—	2
Barley bran	10·0	14·9	3·6	5·0	8·6	57·9	—	3
Barley dust	13·0	11·8	2·2	3·2	4·6	65·2	—	3

* 1. Chatfield and Adams (1940). 2. Original data. 3. Watson (1953).

Vitamins. During malting of barley there is little change in the vitamin B_1 content, but riboflavin increases 2–3 fold, and there are increases of 10–25% in the nicotinic acid content and of 40–50% in the pantothenic acid content (cf. p. 57). Pearl barley contains, per g, about 1·2, 0·35 and 25 μg of vitamin B_1, riboflavin and nicotinic acid, respectively.

REFERENCES

BIRTWISTLE, S. E., HUDSON, J. and McWILLIAM, I. C. (1962), Use of unmalted wheat flour in brewing, *J. Inst. Brewing* **68**: 467.

CHATFIELD, C. and ADAMS, G. (1940), *Food Composition*, U.S. Dept. Agric. Circ. 549.

GEDDES, W. F. (1936), Proximate feeding-stuff analyses of western Canadian barley, oats and rye, *Canada Dom. Grain Research Lab. Ann. Rpt.* **10**: 83.

HILL, D. D. (1933), The chemical composition and grades of barley and oat varieties, *J. Amer. Soc. Agron.* **25**: 301.

HUNTER, H. (1928), *The Barley Crop*, Crosby Lockwood, London.

MORRISON, F. B. (1947), *Feeds and Feeding*, 20th edition, Morrison Publ. Co., Ithaca, N.Y.

PERCIVAL, J. (1921), *The Wheat Plant*, Duckworth, London.

SCHUSTER, K. (1962), Malting technology, in COOKE, A. H. (1962).

WATSON, S. J. (1953), The quality of cereals and their industrial uses. The uses of barley other than malting, *Chem. & Ind.*, p 95.

WRIGHTSON, J. and PATRICK WRIGHT, R. (1908), Barley, in *Standard Cyclopaedia of Modern Agriculture*, Gresham Publ. Co., London.

FURTHER READING

ABBOTT, A. W., Barley meal plants, *Miller* **61**: 1098, 1935.

ANON. Barley meal, *Milling* **83**: 240, 1934.

COOKE, A. H. (Ed.), *Barley and Malt*, Academic Press, New York and London, 1962.

DOUGLAS, A., GRANT, W. and KENT, N. L., *Milling of Barley and Oats in Germany*, B.I.O.S. Final Report No. 718, Item 31, H.M.S.O., London, 1946.

HINTON, J. J. C., Comparative morphology and biology of wheat and barley grains and the location of nutrients, *Recent Advances in the Processing of Cereals*, Soc. Chem. Ind. Monograph No. 16, Soc. Chem. Ind., London, 1962.

KASARYAN, S., Milling and baking experiments with barley, *Northw. Miller* **182** (5): 469, 1935.

KENT, N. L., Some characteristics of barley, *Milling* **101**: 240, 1943.

NATIONAL INSTITUTE OF AGRICULTURAL BOTANY, *Cereal Quality*, Farmers Leaflet No. 14, N.I.A.B., Cambridge, 1963.

CHAPTER 13

OATS:
PROCESSING, NUTRITIONAL
ATTRIBUTES, TECHNOLOGICAL USES[1]

UTILIZATION

The oat crop, which in Britain occupied a slightly larger acreage than wheat until 1960, is used mainly for the feeding of farm animals. A small proportion is milled to provide products for the human dietary: oatmeal for porridge and oatcake baking, rolled oats for porridge, oat flour for baby foods and for the manufacture of ready-to-eat breakfast cereals, and "white groats" for making "black puddings"—a popular dish in the Midlands of England.

Usage of oats for human consumption in 1961–2 amounted to 7·5% of the total home crop in the U.K., 4% in the U.S.A., 2% in Canada, and only 0·2% in France. In all these countries, 6–7·5% was used for seed, the rest for animal feeding. (Source: *Grain Crops*.)

PROCESSING OF OATS

The processing of oats in the mill differs from that of wheat because of the differences in anatomical structure and chemical composition between the two cereals, and because the purposes for which the products are intended are not the same for the two cereals: indeed, the difference in purposes is partly the

[1] Much of the information in this chapter has been collected at the Cereals Research Station, St. Albans, in connection with a programme of work for the Oatmeal Millers' Associations of England, Scotland, and Northern Ireland during the 14 years 1947–60.

consequence of the differences in chemical composition. For example:

1. The adherent husk of the oat grain, like that of the barley grain (cf. p. 205), is tough and fibrous, and quite inedible by the human. It must therefore be removed in a special shelling process during the manufacture of edible products, for which only the kernel (groat) is required. The chaff of wheat, which corresponds morphologically with the husk of oats, comes away easily during threshing.

2. The fat content of the oat kernel is 2–5 times as high as that of wheat (cf. Table 10), and the oat kernel contains an active lipase (fat-splitting) enzyme (cf. p. 49), and possibly other undesirable enzymes. These facts have an important bearing on the storage quality of the products, and the processing of oats must be designed to minimize the possibility of rancidity developing in the products.

3. The protein of the oat kernel, although not less in quantity than that of the wheat grain and, like the latter, composed of amino acids, does not form gluten when mixed with water. Thus, oat flour and oatmeal cannot be used for making bread that compares in light spongy quality with wheaten bread. Accordingly, the processing of oats is designed to make products suitable for porridge-making, oatcake baking, ready-to-eat breakfast cereals, and infant foods.

4. The bran of the oat kernel is relatively thin and pale in colour, in comparison with that of wheat. For most purposes, the presence of the bran has little detrimental effect on the quality of the milled oat products. Oatmeal and rolled oats are manufactured as whole groat meal: that is, the bran is not separated from the endosperm; as a result, the milling process is considerably simplified and the potential yield of milled products is increased. This increase in yield is, of course, offset by the large proportion of husk which is discarded as a by-product. Removal of the pericarp, e.g. by the wet-brushing process (cf. p. 216), is, however, advantageous in the manufacture of infant cereal from oats, as the fibre content can thereby be reduced by 30–40%.

OATMEAL MILLING

The sequence of operations in oatmeal milling is shown in Fig. 33. The oats as received from the farm must be cleaned, removing not only any extraneous matter but also various types of unproductive and undesirable oat grains, e.g. "light" grains or empty husks, and frit-fly-infested grains with powdery discoloured kernels. The processes and machines used in cleaning oats are similar to those described for wheat cleaning (Ch. 6), viz. indented discs or cylinders, aspirators, sieves and electrostatic separators.

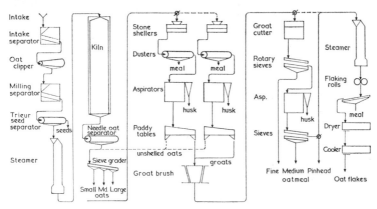

Fig. 33. Flow sheet for milling oatmeal and oatflakes (rolled oats).

The cleaned oats are stabilized in order to inactivate lipase.

Lipase. The enzyme lipase, present in raw oats, is located almost entirely in the pericarp (the outer layers) of the kernel or groat (Hutchinson *et al.*, 1951).

Lipase hydrolyses fats into their component parts. Oat fat is a compound of glycerol and fatty acids, principally oleic, linoleic and palmitic (cf. Table 20). The action of lipase on oat fat is thus to split it into glycerol and free fatty acids (FFA).

Oat fat is distributed throughout the endosperm, germ and aleurone layer, the last two being particularly rich sources. There is little if any fat in the pericarp.

In the intact kernel of raw oats the lipase and fat thus do not come into contact with each other, being located in different parts of the kernel, and hence lipase exerts little or no effect on the fat. Raw sound oats may have quite an appreciable potential lipolytic activity, but so long as the kernels remain entire the lipase is latent, and the FFA content of the kernels does not increase to any marked degree during storage. Sound oat kernels normally contain up to 10% of the fat (6·7% of the fat, on average) in the form of FFA.

When the oat kernels are broken or ground in the process of oatmeal milling, the lipase in the pericarp is brought into contact with the oat fat in the endosperm, germ and aleurone, and the process of FFA release is accelerated to an extent dependent on moisture content, temperature and fineness of grinding. The moisture content is particularly important in this respect: lipase activity is low in oatmeal stored at low moisture content and low temperature, but is appreciable at 10% or higher moisture content.

Free fatty acids. Glycerol is neutral and stable, and when released from oat fat by hydrolysis does not give rise to objectionable effects. The slow release of FFA and related substances from oat fat, however, is accompanied by development of the bitterness associated with samples of milled oat products in which lipase is active.

In the baking of oatcakes from oatmeal, fat and water, the raising agent frequently used is sodium bicarbonate. If FFAs are present in the oatmeal, they react with the bicarbonate to form the sodium salts of fatty acids, which are soaps. The resulting oatcakes thus have a soapy flavour.

The fat added to the oatmeal in oatcake baking may be animal fat, e.g. beef dripping, or vegetable fat. All types of fat consist of glycerol plus fatty acids, the particular fatty acids concerned varying from fat to fat. The use of certain vegetable fats, e.g. palm kernel and coconut oil, in oatcake baking is particularly undesirable as the fatty acids in these fats are chiefly lauric and myristic; the former, when released by lipase, has an unpleasant

soapy taste. A summary of the main points in the relationship
between lipase and fat in oatmeal milling is shown in Fig. 34.

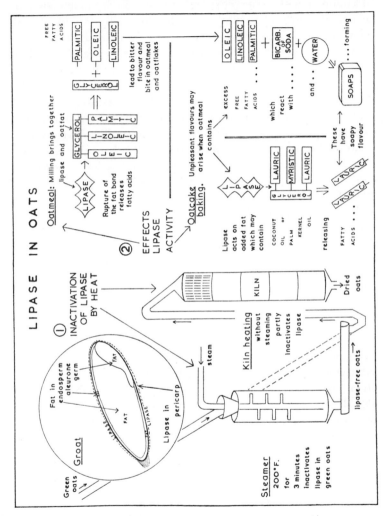

Fig. 34. Main points in the relationship between fat and the enzyme
lipase in oatmeal milling and oatcake baking.

Inactivation of lipase. Lipase serves no useful purpose in milled oat products, and is desirably inactivated before or during milling. In common with other enzymes, lipase can be inactivated by heat treatment under suitable conditions, or by treatment with acids. Inactivation by heat treatment is more rapid at high moisture content than at low; the best results are obtained by treating the raw or green oats with steam, before kiln-drying, in the process known as stabilization. The stabilization process was introduced largely on account of the undesirably high FFA content frequently found in oatmeal milled from unsteamed oats; rolled oats (see below) do not normally develop FFA because the process of steaming the pinhead meal is generally adequate to inactivate lipase. In the stabilization process, the oats, at 14–20% m.c., are quickly raised to a temperature of 205–212°F by injection of live steam at atmospheric pressure, and thereafter maintained at that temperature for 2–3 min by controlling the rate of throughput of the steamer. The stabilization process, besides inactivating lipase and most of the other unwanted enzymes present, appears to have a beneficial effect on development of flavour, and on resistance to the onset of oxidative rancidity.

A check on the completeness of lipase inactivation may be made by applying the tetrazolium test (cf. p. 91). If, after stabilization, a red colour develops at the germ end of the kernel, as with living grain, it follows that the heat treatment used was insufficient to inactivate the dehydrogenase enzymes, and that other enzymes—in particular, lipase—have probably survived also. If the oats show residual lipase activity after stabilization, the rate of steamer throughput can be reduced (giving a longer treatment), or the steam pressure can be increased (giving more rapid heating). Excessive steaming beyond that required for enzyme inactivation is to be avoided, because oxidation of the fat, resulting in oxidative rancidity (cf. p. 49), may be encouraged by heat treatment.

An alternative method of dealing with lipase is a "wet-brushing" process developed by Martin (1954), in which groats are vigorously scrubbed while immersed in water or dilute acid (e.g. decinormal HCl). About 95% of the pericarp (and the lipase with it) is removed by this treatment, only that portion buried deep in the crease being inaccessible. Removal of the pericarp also improves the colour (the pericarp being the coloured layer in groats) and reduces the fibre content of the milled products.

Kiln drying. After stabilization, the oats are dried to 4–8% m.c. The old traditional method made use of a flat-head kiln (Fig. 35) consisting of a square room floored with perforated metal sheeting, upon which the oats were spread to a depth of some 4 in. Hot furnace gases were led up through the perforated floor, and the oats were turned over with wooden shovels periodically

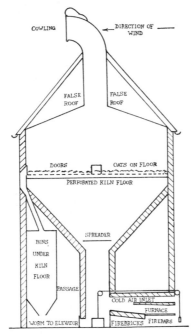

Fig. 35. Flathead kiln as used in some Scottish oatmeal mills. (Drawing by J. Thomlinson.)

during the kilning time of about 4 hr in an attempt to obtain uniform treatment. Flat-head kilning was a batch process, with many disadvantages, but produced oats with flavour rarely matched by other processes. Continuous drying is nowadays frequently carried out in a Walworth kiln (Fig. 36) in which the source of heat is the hot air, carrying the volatile products of combustion, from a furnace fired with coke or anthracite.

The purposes of kiln-drying are: (1) to reduce the moisture content to a level that is satisfactory for the storage of the milled products; (2) to facilitate the subsequent shelling of the oats by increasing the brittleness of the husk; and (3) to develop in the

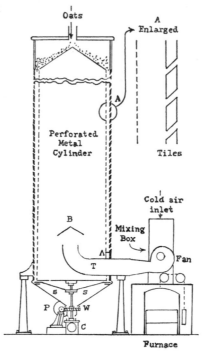

Fig. 36. Vertical section through a Walworth kiln used for drying oats. (Drawing by J. Thomlinson. Reproduced from N. L. Kent, *Cereal Sci. Today* **2**: 83, 1957, by courtesy of the Editor.)

oats a characteristic flavour which is often described as "nutty". Before the introduction of stabilization (about 1950) the kilning process was also relied upon to reduce the lipolytic activity. It is improbable that complete inactivation was achieved: the necessary temperature required to inactivate 97% of the lipase in oats increases from 148°F at 20% m.c. to 180°F at 12% m.c., and to 219°F at 8% m.c. During kilning, the temperature of the oats often exceeds 195°F, but by the time this temperature has been reached the moisture content has fallen below 12%, so that the necessary conditions for complete inactivation are not attained.

Development of flavour in the oat during kilning was investigated by Moran *et al.* (1954). Optimum conditions for flavour development consisted in the gentle drying of the grain to about 8% m.c. at a grain temperature not exceeding 175°F, followed by a short toasting for about 20 min in a current of air at 300°F to reduce the moisture content to about 5%. Flavour development was not related to variations in chemical composition of the oats, and was not due to volatiles derived from the fuel, nor to any constituent derived from the husk. Stabilization did not affect the development of flavour during subsequent kilning. The uniformity of treatment of every grain in the kiln is an important requirement, but difficult to achieve in practice. Under-kilned grains contribute no flavour, whilst over-kilned grains contribute an undesirable flavour and may ruin the whole parcel.

Dry-shelling. In the conventional dry-shelling process, the kiln-dried oats are passed between a pair of large circular stones, one of which is stationary, the other revolving. The two stones are separated by a distance slightly less than the length of the oat grain: as the grains roll over and up-end the husk is split off in thin slivers. Careful adjustment of the sheller and grading of the oats before shelling, so that the feed consists of grains of uniform length, ensures minimum breakage of the kernels. The mixture of kernels, husk slivers and the few unshelled oats is aspirated by air currents, which lift the husk away from the groats and unshelled oats. These two constituents of the mixture are then separated from each other on a "paddy" machine, or inclined table separator, which causes the groats and the oats to move in opposite directions by taking advantage of the difference in specific gravity and in resiliency (or "bounce") between oats and groats. The unshelled oats are returned to the shelling stage.

Other shelling methods are mentioned below (pp. 221–2).

Polishing and cutting. The shelled groats are brushed or scoured to detach the fine hairs that cover much of their surface. The hairs, known as "dannack", are removed by sieving and aspiration, and become part of the by-product called "oat dust" (see p. 224). The brushed groats are cut transversely, generally by a drum cutter, each kernel yielding four or five pieces called "pinhead meal", of average weight about 6 mg per particle. The cutting process produces a small amount of flour, which is sieved off; this is known as "flow meal" and is used for making dog biscuits.

Grinding. From the pinhead meal, finer cuts of oatmeal ("medium", "fine") are made by grinding on stones and then sifting. The average yield of oatmeal is 57%, by-products accounting for 28%, and loss (mostly moisture) for 15%.

Manufacture of rolled oats. Oat flakes or rolled oats for quick-cooking porridge are made from pinhead meal (or sometimes from whole uncut groats) by cooking the pinhead meal in a steamer, rolling the cooked product while hot, moist and plastic, between heavy rollers, and drying the flakes so formed. The flakes are cooled by a current of cold air and sifted to remove floury particles before packing. Oatmeal is packed at 8–10% m.c., rolled oats at about 10·5% m.c.

The amount of domestic cooking required by rolled oats in porridge making is dependent to a large extent on the processes of cutting, steaming and flaking, which are interrelated. The size of the pinhead meal influences rate of moisture penetration in the steamer; smaller particles will be more thoroughly moistened than large particles by the steaming process, and hence the starch will be gelatinized to a greater degree, and the steamed pinhead meal will be softer. For a given roll pressure at the flaking stage, this increase in softness will result in thinner flakes being obtained from smaller-sized particles of pinhead meal. During the cooking of porridge, the thinner flakes will cook more rapidly than thicker flakes because moisture penetration is more rapid.

Thin flakes would normally be more fragile than thick ones, and more likely to break during transit. However, thin flakes can be strengthened by raising the moisture content of the pinhead meal feeding the steamer, because the degree of gelatinization of the starch is thereby increased. Gelatinized starch has an adhesive quality, and quite thin flakes rolled from highly gelatinized small pinhead meal can be surprisingly strong.

Commercial rolled oat flakes are generally 0·012–0·015 in. in average thickness; when tested with Congo Red stain (which colours only the gelatinized and damaged granules, cf. p. 175), about 30% of the starch granules appear to be gelatinized.

Green-shelling process. In the conventional dry-shelling process described above, which is in use in the majority of oatmeal mills in Britain, the oats are shelled at a relatively low moisture content (about 6%) after kiln drying. In the green-shelling process, the oats, after stabilization by steaming (see above), are shelled at natural moisture content, viz. 14–18%, on impact hullers such as the Murmac huller, in which the oats are caused to impinge at high velocity against a hard plate faced with abrasive material. The husks are separated from the groats by aspiration in the normal way, and the *groats* are then kiln-dried. Herein is a complication of the green-shelling process, because the high temperatures in the later part of the process that have been found necessary for the development of flavour during the kilning of *oats* (cf. p. 217) are quite unsuitable for the kiln-drying of groats; the groats no longer have the protection afforded by the husk in the case of the whole oats, and become burnt and discoloured at the high temperature. Moreover, high temperature treatment of the groats at low moisture content may lead to the onset of oxidative rancidity. As a result of these limitations, groats are kiln-dried at somewhat lower temperatures than those used for whole oats, and the products are devoid of the typical oaten flavour. The flavour of oaten products made by the green-shelling process, often described as "bland", seems insipid to palates familiar with products made by the conventional process.

The high temperature kilning of oats facilitates the subsequent inactivation of residual lipase by steam in the cooking process to which pinhead meal is subjected before flaking, whereas more severe steam treatment is necessary to inactivate lipase in the case of products made by the green-shelling process, in which lower kilning temperatures are used.

Wet-shelling process. This process, invented by Hamring (1950), is basically similar to the green-shelling process in being an impact shelling method, but differs from it in that the oats are first damped to 22% or higher moisture content before shelling. The mixture of groats, shells and unshelled oats is then dried before the components are separated from each other. It is claimed that the preliminary moistening decreases the breakages of groats and increases the efficiency of shelling, in comparison with green-shelling, but tests have shown that the efficiency of shelling did not exceed about 90%.

American system of kilning. Oatmeal millers in the U.S.A. use a batch method of kilning which is intermediate between the flathead and the Walworth kilning processes. The oats are heated, to a set temperature and for a set time, in a pan with constant stirring. In some cases the oats cascade through a series of pans heated to progressively higher temperatures. The product from this method is said to be more uniform than that from the Walworth kiln.

White groats. This product, used for black puddings and haggis, is made by damping groats and subjecting them to a vigorous scouring action in a barley polisher, pearler or blocker before the moisture has penetrated deeply into the kernels. Alternatively, the groats may be scoured without any preliminary damping. A proportion of the pericarp is removed by this process. There is also a certain amount of breakage, with the release of oat flour: the latter becomes pasted over the groats (no attempt being made to remove it), which thus acquire a whitened appearance. The pericarp is removed much less efficiently by this process, as practised in the mill, than by the wet-brushing process (cf. p. 216).

NUTRITIVE VALUE

The chemical composition of milled oat products is shown in Table 49.

TABLE 49
CHEMICAL COMPOSITION OF MILLED OAT PRODUCTS

Material	Moisture (%)	Protein (N × 6·25) (%)	Fat (%)	Ash (%)	Fibre (%)	Carbo-hydrate (%)
Oatmeal	8·8	13·0	6·8	1·9	1·1	68·4
Rolled oats	10·1	13·2	6·8	1·8	0·9	67·2
Oat flour	9·3	14·1	7·2	1·8	1·0	66·6

The vitamin B_1 content of oatmeal is about 5 $\mu g/g$ (McCance and Widdowson, 1960); Holman and Godden (1947) give a range from 3·6 to 6·0 (mean: 4·7) $\mu g/g$ for vitamin B_1 content of kernels of Scottish-grown oats. Higher values (6·7 $\mu g/g$ in whole oats, equivalent to 8·7 $\mu g/g$ in the kernel) are reported in Canadian-grown oats (Robinson *et al.*, 1950). Data for niacin and riboflavin contents of oats and oat products are shown in Table 50.

TABLE 50
NIACIN AND RIBOFLAVIN CONTENTS OF OATS

Material	Niacin ($\mu g/g$)	Riboflavin ($\mu g/g$)	Source of data*
Whole oats	6·8	1·3	1
Oat kernels	7·2	1·5	1
Oat flour	7·2	1·1	2
Oat husk	4·5	1·0	1

*1. Shaw (1953); 2. Hellstrom and Andersson (1953).

Daniels and Martin (1961) isolated naturally occurring anti-oxidants from oats. The substances appear to be complexes of caffeic and ferulic acids joined by a long chain hydroxy fatty acid.

The antioxidant activity of individual members of the group is related to their caffeic acid content, and is of the same order as that of the synthetic antioxidants BHT and propyl gallate (Daniels *et al.*, 1963) (cf. p. 196).

BY-PRODUCTS OF OATMEAL MILLING

The main by-products of oatmeal milling are oat husk, oat dust and meal seeds, which constitute about 70, 20 and 10%, respectively, of the total milling by-products. Typical figures for their chemical composition are shown in Table 51.

Commercial oat husk consists principally of true (botanical) husk, but generally has an admixture of up to 10% of kernel material.

TABLE 51

CHEMICAL COMPOSITION OF BY-PRODUCTS OF OATMEAL MILLING

Material	Moisture (%)	Protein* (%)	Fat (%)	Ash (%)	Fibre (%)	Starch (%)	Undetermined (%)
Oat husk	4·6	1·3	0·4	4·3	36·1	0·9	52·4
Oat dust	5·4	10·0	4·5	6·0	21·6	10·0	42·5
Meal seeds	8·2	7·9	3·5	2·8	17·2	26·4	34·0
Oat feed meal†	8·0	3·4	1·5	3·7	30·0	5·7	47·7

* N × 6·25.   † Oatmill feed in the U.S.A.

Oat dust, comprising about six parts of "light dust" to one part of "heavy dust", is rich in kernel material, particularly the outer layers of the kernel, and contains about 16% by weight of groat hairs (dannack), and about 11% of husk fragments. The "heavy dust", with a fat content of about 9%, is relatively richer in kernel and poorer in husk than the "light dust".

Meal seeds consists of slivers of husk and fragments of kernel in approximately equal proportions.

Utilization. Oat dust and meal seeds are of reasonably good feeding value and are used for animal feeding.

Meal seeds is also used for making sowens, a peculiar Scottish dish. The seeds is steeped in water and allowed to ferment for several days until it becomes acid. The liquid is strained off through muslin, discarding the solids, including the husk.

Before 1939, oat husk was burned as fuel to fire oat kilns. During the war, it was ground to make oat husk meal or, when admixed with other by-products of oatmeal milling, oat feed meal (oatmill feed in the U.S.A.), a feed of low nutritive value for ruminant animals.

Oat husk is used industrially as a raw material in the manufacture of furfural, which is obtained from it in 10–14% yield. Plants for this purpose have been operated in the U.S.A. at Cedar Rapids, Iowa; Memphis, Tenn.; and Omaha, Neb. Oat husk is also used as a filter-aid in breweries, where it is mixed with the ground malt and water in the mash tun (cf. p. 203) in order to keep the mass porous; as a diluent or filler in linoleum; as deep litter for battery chickens; and as an abrasive in air-blasting for removing oil and products of corrosion from machined metal components.

REFERENCES

DANIELS, D. G. H., KING, H. G. C. and MARTIN, H. F. (1963), Antioxidants in oats: esters of phenolic acids, *J. Sci. Fd. Agric.* **14**: 385.

DANIELS, D. G. H. and MARTIN, H. F. (1961), Isolation of a new antioxidant from oats, *Nature, Lond.* **191**: 1302.

HAMRING, E. (1950), Wet shelling process for oats, *Getreide Mehl Brot* **4**: 177.

HELLSTROM, V. and ANDERSSON, R. (1953), Content of B vitamins in oats and oat flour, *Var. Föda* **5** (10): 41.

HOLMAN, W. I. M. and GODDEN, W. (1947), The aneurin (vitamin B_1) content of oats, *J. Agric. Sci.* **37**: 51.

HUTCHINSON, J. B., MARTIN, H. F. and MORAN, T. (1951), Location and destruction of lipase in oats, *Nature, Lond.* **167**: 758.

McCANCE, R. A. and WIDDOWSON, E. M. (1960), *The Composition of Foods*, Med. Res. Coun., Spec. Rep. Ser. 297, H.M.S.O., London.

MARTIN, H. F. (1954), Improvement of cereal grains, Brit. Pat. Spec. 715,943.

MORAN, T., HUTCHINSON, J. B. and THOMLINSON, J. (1954), The flavour of porridge, *Nature, Lond.* **174**: 458.

ROBINSON, A. D., TOBIAS, C. H. and MILES, B. J. (1950), The thiamine and riboflavin content of Manitoba grown wheat, oats and barley of the 1947 crop, *Canad. J. Res.* **28F**: 341.

SHAW, B. (1953), Unpublished data.

FURTHER READING

BERRY R. A., Composition and properties of oat grain and straw, *J. Agric. Sci.* **10**: 359 1920.

BRIT. PAT. SPEC. Nos. 819,241 (kilning); 583,188; 626,883; 659,950 (hulling); 585,772 (furfural); 647,085 (stabilization); 809,962; 844,657 (electrostatic separation).

BROWN, I., SYMONS, E. F. and WILSON, B. W., Furfural: a pilot plant investigation of its production from Australian raw materials, *J. Coun. Sci. Industr. Res.* **20**: 225, 1947.

BROWNLEE, H. J. and GUNDERSON, G. L., Oats and oat products, *Cereal Chem.* **15**: 257, 1938.

COFFMAN, F. A., *Oats and Oat Improvement*, Amer. Soc. Agron., Madison, Wis., 1961.

DEPARTMENT OF AGRICULTURE FOR SCOTLAND, *Oat Varieties*, Leaflet No. 1, H.M.S.O., Edinburgh, 1948.

FAGAN, T. W., Oats, their milling and by-products, *Scot. J. Agric.* **11**: 307, 1919.

FINDLAY, W. M., *Oats*, Oliver & Boyd, Edinburgh, 1956.

HITCHCOCK, L. B. and DUFFEY, H. R., Commercial production of furfural in its 25th year, *Chem. Engng. Prog.* **44**: 669, 1948.

HUNTER, H., *Oats*, Benn, London, 1924.

HUTCHINSON, J. B., The quality of cereals and their industrial uses, *Chem. & Ind.*, p. 578, 1953.

HUTCHINSON, J. B. and MARTIN, H. F., The chemical composition of oats, *J. Agric. Sci.* **45**: 411, 419, 1955.

KENT, N. L., The quality of oat products, *Milling* **126**: 586, 1956.

KENT, N. L., Recent research on oatmeal, *Cereal Sci. Today* **2**: 83, 1957.

KENT, N. L., The food value of oats, *Health for All* **34**: 291, 1960.

LEA, P., A simple outline of oatmeal milling, *J. Inst. Corn Merch.* **1**: 8, 43, 88, 1947.

McCANCE, R. A. and GLASER, E. M., Energy value of oatmeal and digestibility and absorption of its proteins, fats and calcium, *Brit. J. Nutr.* **2**: 221, 1948.

NATIONAL INSTITUTE OF AGRICULTURAL BOTANY, *Varieties of Oats*, Farmers Leaflet No. 13, N.I.A.B., Cambridge, 1964 (issued annually).

SCIENCE EDITOR, Oat products: their nutritive and dietary value, *Milling* **120**: 643, 1953.

SPISNI, D., Chemical composition and biological value of oats, *Clin. Vet.* **73**: 353, 1951.

WOODMAN, H. E., Notes on feeding (oatmilling by-products), *Agriculture* **48**: 159, 1941.

CHAPTER 14

RYE, RICE, MAIZE: PROCESSING, NUTRITIONAL ATTRIBUTES, TECHNOLOGICAL USES

RYE

Utilization

Rye (*Secale cereale*) is a bread grain, second only to wheat in importance, and it is the main bread grain of Scandinavian and eastern European countries. Although nutritious, and palatable to some people, rye bread is not comparable with wheaten bread as regards crumb quality and bold appearance of the loaf; as living standards rise, the consumption of rye bread falls, while that of wheaten bread rises. The production of rye exceeded that of wheat in Western Germany from 1939 until 1957; thereafter, production declined, and since 1958 a larger amount of rye has been used for animal feed than for human food. Production of rye, however, still exceeds that of wheat in Eastern Germany. Outside Europe, rye is used mainly for animal feed. A small amount is used for distilling (cf. p. 204).[1]

Of the total annual domestic consumption of 3·4 million tons of rye in Western Germany in 1961–2, 41% was used for human food, 51% for animal feed, 5·5% for seed and 1·7% for industrial purposes. Consumption of rye in the U.S.A. was only 0·6 million tons in the season 1962–3; 34% was used for animal feed, 21% for human food, 27% for seed, and 17% for industrial purposes.[1]

[1] *Grain Crops*. See footnote on p. 1.

Historical

Rye was domesticated relatively recently, about the fourth century B.C. in Germany, later in southern Europe. According to Vavilov (1926), cultivated rye has been derived from the rye grass that occurred as a weed in wheat and barley crops.

In Roman times the chief cereal crop in the south was probably wheat, but rye was introduced by Teutonic invaders, who used it for making bread, and it was grown in East Anglia.

During the Middle Ages the poorer people in England ate bread made from rye, or from a mixture of rye and wheat known as maslin, or from barley and rye (cf. p. 198). In 1764, according to Ashley (1928), bread made in the north of England contained 30% of rye, that in Wales 40%. At this time, rye was still an important crop in the north of the country and, in addition, was regularly imported from Germany and Poland, where it was plentifully grown.

Rye milling

The products of rye milling are rye flour of various extraction rates for soft breads, coarse rye meal for hard breads, and rye flakes for hot breakfast cereals.

The processes of rye milling, like those of wheat milling, comprise cleaning, conditioning, and the milling process proper.

The cleaning process utilizes principles and machinery essentially similar to those used in wheat cleaning (cf. Ch. 6). There are two complications with rye:

1. The grain is variable in size, weighing from 15 g to 40 g per 1000 grains. Accordingly, care must be exercised, for example, by selecting suitably sized sieves, that a considerable proportion of small or large grains is not rejected with the screenings. The newer varieties of rye are more uniform in grain size.

2. The grain is liable to contamination with ergot (cf. p. 232), which should be separated before the grain reaches machines, such as scourers, that might break up the ergots and render their removal more difficult. The milling separator (used with

suitable sieves), indented cylinders or discs, and table separators are used for rye cleaning; in addition, rye is washed in a washer and whizzer as used for wheat washing.

Ergots, being generally of greater length than rye grains (cf. p. 233), are separated on the milling separators or the indented discs. They can also be separated by flotation in strong salt solution (sp. gr. 1·12) in which the rye grains sink while the ergots float (Plante and Sutherland, 1944). In order to recover the maximum amount of ergots for medicinal purposes from a rye–ergot mixture, these workers recommended treating the mixture with a dilute paraffin-oil emulsion. The paraffin spreads over the grains but not over the ergots: as a result, the grains float in water and can be skimmed off, while the ergots sink.

In the conditioning process, rye is brought to a suitable moisture content for milling; British and European ryes are dried to 14–15% m.c., but the stronger Canadian and Australian ryes are damped and hot conditioned.

Milling. Rye is milled by grinding the cleaned and conditioned grain on a succession of pairs of rolls, all or the majority of which are fluted, and sifting out the flour from the grind. In a general way, the process resembles that of wheat flour milling (cf. Ch. 7) but diverges from the latter in detail because of two important points of difference between rye and wheat in their manner of breakdown: (1) the endosperm of rye comminutes to flour fineness more readily than does that of wheat, but the flour is difficult to sift; (2) the endosperm of rye parts from the bran more reluctantly than does that of wheat.

In the milling process adopted for rye, a large amount of flour is made on the early grinds (breaks), but little semolina is recovered. This differs from wheat milling, where the principal product of the break milling is semolina. Each break grind in rye milling is scalped (cf. p. 115), the flour dressed out, and the overtails fed to the next break. Fractions of intermediate particle size in the I and II Bk. grinds are reduced on fluted or smooth middlings reduction rolls.

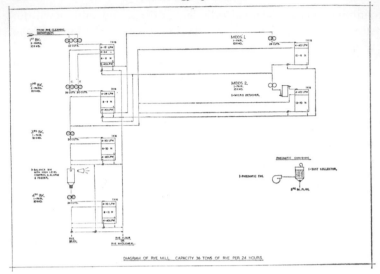

Fig. 37. Diagram of a rye mill flow, suitable for Scandinavian conditions, from which a variety of grades of flour and meal can be produced. (By courtesy of Thomas Robinson & Son Ltd.)

The diagram of a rye milling flow is shown in Fig. 37. The Break flutings and the scalping covers become progressively finer from 1st to 4th Bk. Plansifters are used for sifting. The absence of purifiers is notable. A recent development is the use of impact grinding in combination with rollermilling: the diagram shows a micro-detacher taking the grind from Midds 2 prior to sifting. The milling system can by this means be shortened, and a flour less contaminated with bran obtained.

The yield of reasonably pure flour from rye is 64–65%. At increasing extraction rates the flour becomes progressively darker in colour and more fibrous, and the characteristic rye flavour more prominent. Rye meal may be of any extraction rate, but rye wholemeal is 100% extraction.

The lighter grades of rye flour are bleached, generally with chlorine, although nitrogen peroxide and benzoyl peroxide are sometimes used. The chlorine does not "improve" the baking quality, as it does in the case of wheat flour (cf. p. 138). Chlorine dioxide is likewise ineffecitve because of the nature of the rye protein. Whole-grain rye is sometimes bleached with sulphur dioxide, prior to milling, in European countries, but not in Britain or the U.S.A. (cf. p. 206).

Rye cones is a by-product of rye milling, consisting of dusted fine and medium semolina, together with small particles of germ and bran, all passing through a No. 40 wire sieve (aperture: 0·47 mm). It is mixed back with short extraction rye flour to make flours or meals of longer extraction rates.

Nutritive value

The chemical composition of various increments in commercially milled 73% extraction rye flour and of the milling offals, and of rye flour laboratory-milled to various extraction rates is shown in Tables 52 and 53.

TABLE 52

CHEMICAL COMPOSITION OF RYE AND MILLED RYE PRODUCTS

(15% M.C. BASIS)

Material	Protein (%)	Fat (%)	Ash (%)	Fibre (%)	Carbo-hydrate (%)	Energy (calories per 100 g)	Source of data*
Rye grain	9·9	1·6	1·7	1·7	70·2	334	1
Rye flour							
0–31% increment	5·7	0·6	0·4	0·1	78·2	341	1
31–62% increment	9·3	1·2	0·8	0·3	73·3	341	1
62–67% increment	12·3	2·0	1·4	0·8	68·5	341	1
67–73% increment	14·1	2·3	1·8	1·0	65·8	340	1
Rye milling offals†							
Bran, fine	15·0	3·1	4·1	4·9	58·0	—	1
Bran, coarse	17·8	5·1	3·7	9·2	49·2	—	1
Germ	38·0	10·1	4·7	3·3	37·4	—	1
Rye flour:							
60% extraction	5·6	1·0	0·5	0·2	78	346	2
75% extraction	6·7	1·3	0·7	0·5	75	339	2
85% extraction	7·3	1·6	1·0	0·8	73	336	2
100% extraction	8·0	2·0	1·7	1·6	69	326	2

* 1. Neumann *et al.* (1913). 2. McCance *et al.* (1945).

† Offals remaining after milling of 73% extraction flour: yield of fine bran 23·4%, coarse bran 2·4%, germ 0·2%.

TABLE 53

VITAMIN AND MINERAL CONTENTS OF RYE FLOURS*

(15% m.c. basis)

Extraction	K (mg/ 100 g)	Ca (mg/ 100 g)	Mg (mg/ 100 g)	Fe (mg/ 100 g)	Total P (mg/ 100 g)	Phytate P (mg/ 100 g)	Vitamin B$_1$ (μg/g)	Ribo-flavin (μg/g)
60%	140	15·3	16	1·3	78	24	1·5	0·8
75%	172	19·5	26	1·7	129	57	2·6	1·4
85%	203	26·1	45	2·0	193	104	3·2	2·0
100%	412	31·5	92	2·7	359	258	4·6	2·9

* Source: McCance *et al.* (1945).

Somewhat higher values for protein, ash and fibre are given by Schopmeyer (1962) for rye flours milled in the U.S.A.

In comparison with wheat milling, the milling of rye removes proportionately more protein from the grain, particularly at the lower extractions; on the other hand, separation of fibre is less complete than in wheat milling.

The nicotinic acid (niacin) content of 60% rye flour is about 9 μg/g.

Ergot and ergotism

Ergot is the name given to the sclerotia of the fungus *Claviceps purpurea* which infects many species of grasses and is particularly liable to infect rye (see Fig. 38). Wheat, barley and oats are also attacked, but comparatively rarely.

Ergot has been associated with rye because the latter was generally grown on soils which were too poor to give a useful crop of other cereals, but which provided suitable conditions for *Claviceps*. Rye grown on good land, from fresh seed, is probably no more liable than wheat or barley to become ergotized.

The importance of ergot as a contaminant of rye lies in its toxicity: when consumed in quantity it causes gangrenous ergotism, a disease which was known as "Holy Fire" in the eleventh to sixteenth centuries—although its connection with ergot was not then known.

FIG. 38. Spikes of rye showing ergot sclerotia. (Photo by W. C. Moore. Reproduced from Ministry of Agriculture, Fisheries and Food Bulletin, No. 129, *Cereal Diseases*, by permission of the Controller of H.M.S.O.)

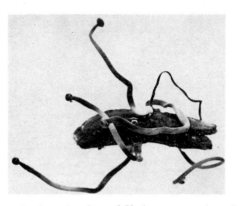

Fig. 39. A germinating sclerotium of *Claviceps purpurea* (ergot), with six stroma. Each consists of a curved stalk surmounted by a capitulum. The capitula, when ripe, discharge ascospores which germinate if they land on the exposed stigma of a rye flower. (Photo by W. C. Moore. Reproduced from Ministry of Agriculture, Fisheries and Food Bulletin, No. 129, *Cereal Diseases*, by permission of the Controller of H.M.S.O.)

The sclerotia of *Claviceps*, average length 14·6 mm and thickness 6·5 mm, are brittle when dry, dull greyish- or purple-black in colour on the outside, dull pinkish-white within (see Fig. 39). They consist of a pseudoparenchyma of closely matted fungal hyphae.

Ergotism is of two types: gangrenous and convulsive.

Gangrenous ergotism, occurring mainly in areas west of the Rhine and particularly in the Sologne district of France, appeared on a large scale in the Middle Ages. It is caused by consumption of ergot, usually present as a contaminant of bread to the extent of 6% or more. The disease is symptomized by swelling and inflammation of a limb, followed by formication (a feeling as of ants crawling over the skin) and violent burning pains (hence the name "Holy Fire"). Later the limb becomes numb, blackened, shrunken and eventually dry and mummified. Sometimes the affected limb separates from the body spontaneously at a joint without pain or loss of blood, and the patient may then recover. Alternatively, the gangrene may spread upwards until the whole body becomes emaciated.

Convulsive ergotism, occurring mainly in areas east of the Rhine, is due to consumption of ergot in conjunction with vitamin A deficiency (Mellanby, 1930). In these areas, newly harvested rye, rich in ergot, would be eaten by poor people who, having no other available food, would be deficient in vitamin A. The symptoms of the disease are fatigue, giddiness, diarrhoea and vomiting, numbness and tingling of the extremities and, in severe cases, convulsions. After the convulsive fit, hunger follows, resulting from the hypoglycaemia due to ergotoxine poisoning. In more recent epidemics in Hanover (Germany) the government has exchanged sound grain for ergoted grain.

Ergot in medicine. Ergot is used in medicine in connection with midwifery as a means of inducing pains in the womb, the first reference to the active drug appearing in a German book of 1582. It was introduced into American official medicine in the early nineteenth century as a means of quickening childbirth.

Ergot in flour. There was a mild epidemic of ergotism in Manchester, England, in 1927 among Jewish immigrants from central Europe who lived on rye bread. This bread was made from flour reported to contain 0·1–0·3% of ergot. An Austrian Official Codex (1911) states that 0·2% of ergot is harmful to health. The U.S.S.R. in 1926 fixed 0·15% as the maximum harmless quantity of ergot in flour. Flour produced in Germany and in the U.S.A. sometimes contains 0·1% of ergot, and objection has not been made to this concentration. The safe limit of ergot in flour would appear to be 0·1–0·15%. With the daily consumption of 1 lb of bread made from flour containing this concentration of ergot, the intake of ergot would be 0·5–0·75 g per day, well below that usually prescribed medicinally, but, of course, continued over a long period.

The principal pharmacological constituents of ergot are the alkaloids ergotinine, $C_{35}H_{39}O_5N_5$, and ergotoxine, $C_{35}H_{41}O_6N_5$.

For a detailed account of ergotism in history and literature, of the chemistry of ergot, and of its use in pharmacology and medicine, see Barger's monograph (1931).

Baking quality of rye

Under certain conditions rye grains germinate while in the harvest field, and then exhibit an increased enzymic activity which may be undesirable for breadmaking purposes. Rye flour with high maltose figure (e.g. 3·5) and low amylograph value (350 or less) is of poor baking quality.

Rye bread. Rye is used for making ordinary soft bread and also for crisp breads which, in Sweden, appear at most meals.

Soft breads, generally consisting of a blend of rye and wheat flours, are made from doughs fermented with sour dough or leaven. In the sour dough process, a starter sour dough is prepared by allowing a rye dough to stand at 75–80°F for several hours to induce a natural lactic acid fermentation caused by grain micro-organisms. Alternatively, rye dough is inoculated with sour milk and rested for a few hours, after which a mixture of acids (acetic, lactic, fumaric) is added to simulate the flavour of a normally soured dough. This sour dough is then mixed with strong flour and yeast to make mild sour rye bread.

Rye crisp bread is generally made from rye wholemeal or from flaked rye, and may be fermented or unfermented. The traditional method used in Sweden is to mix rye meal with snow or powdered ice; expansion of the small air bubbles in the ice-cold foam raises the dough when it is placed in the oven.

Ryvita is a crispbread made from lightly salted wholemeal rye. Rye contains pentosans in greater quantities than in other cereals, and is used in slimming diets because (1) the pentosans gelatinize and swell in the stomach, giving a feeling of satisfaction; (2) hydrolysis of the polysaccharides is slow: blood sugar level rises slowly but is maintained for 5–6 hr, thereby controlling appetite. The bran content of rye crispbread gives it a mild laxative action.

Pumpernickel is a type of soft rye bread made from very coarse rye meal.

Rye flour is characteristically weak (cf. p. 68), but is greatly improved in strength and mixing tolerance by addition of as little as 25% of wheat flour. Bread made from equal quantities of rye and wheat flours ("maslin": cf. p. 228) is soft and white, and is extensively eaten in Germany. Rye bread keeps moist longer than wheaten bread does.

Other uses of rye

Rye is used for making malt and malt flour of good quality and, in Canada and the U.S.A., for the distilling of rye whisky, which requires 100% rye malt.

Rye flour is used as a filler for sauces, soups and custard powders, and in pancake flour in the U.S.A.

Rye starch of 70% extraction is used as a main ingredient of adhesives.

Rye is flaked in the U.S.A. to make a hot breakfast cereal.

Rye flour can be fractionated by air-classification (cf. p. 145);

a flour of 8·5% protein yielded high and low protein fractions of 14·4 and 7·3% protein contents, respectively.

RICE

Utilization

Rice (*Oryza sativa*) is used mainly for human food. In the U.S.A., where the annual domestic consumption is about 0·9 million tons of milled rice per annum, about 75% is used for human food, 9% for animal feed and seed, and 16% for industrial purposes (*Grain Crops*). Corresponding details for the Asian countries, where the total consumption is much greater, are not available. More than half of the rice crop never leaves the farm upon which it is grown.

Drying of rice

Combine-harvested rice generally has a moisture content of about 20% (wet basis) and the grain must be dried immediately to 13–14% m.c. for storage. The grain is dried in continuous-flow driers somewhat resembling the Walworth kiln described for the drying of oats (cf. p. 218).

As rice is consumed mostly in the form of whole unbroken grains, the processing, including the drying, of the *paddy*, or *rough rice* (the threshed grains with adherent hulls), is designed so that a high yield of unbroken grain may be obtained. The total yield of processed rice is made up of the "head yield", viz. the yield of unbroken kernels, plus the yield of broken kernels, obtained from 100 lb of dried rough rice. The market value of whole kernels is greater than that of broken kernels, and it is important, when drying, to avoid conditions that promote breakage. This is a complication not encountered in the drying of other cereal grains, although somewhat resembling the requirements for the drying of macaroni (cf. p. 194).

During the drying process, water moves from the inside to the outside of the kernels; shrinkage occurs, but the outer part shrinks more than the inner. If the rate of drying is too rapid, the stresses set up in the kernel due to uneven shrinkage cause cracking (checking) and fragmentation of the kernel. Recommendations for the operating conditions of rice driers, with these requirements in mind, have been made by Wasserman *et al.* (1957).

Rice milling

The commercial milling of rice comprises cleaning, shelling or hulling (removal of the hulls), and "milling"—a process in which the bran and germ are partially or wholly removed by an abrasive scouring or pearling process.

Some of the machines used in rice cleaning are similar to those used for the cleaning of other cereal grains; a machine used specially for rice is a bearder, which removes beards and stems of the rice plant.

Various types of huller are in use. A machine consisting of two stone discs is similar in construction and operation to that described for oat milling (cf. p. 219). Close grading of the grain by length is an important preliminary to hulling on this type of machine: grains longer than the average are shattered, reducing the yield of head grain, whilst short grains pass through the huller with the hull still intact.

In another type of huller, the grains pass on a horizontal endless rubber band beneath a rubber roller which produces a shearing action on the grains: the rubber band tends to hold the grain while the roller tends to pull the hull away.

The grain issuing from the hulling machine is sifted and aspirated to remove loosened hulls and dust, and is then passed over a paddy or table separator (cf. p. 219) which separates the kernels from the paddy grains which still retain their hull. The latter are returned to the hulling machines. The whole kernels from which the hulls have been removed are known as *brown rice.*

The bran of the brown rice grains is next removed by a machine called, confusingly, a "huller", adapted from a coffee huller (whence the name). It consists of a grooved tapering drum revolving concentrically within a tapered cylinder; the brown rice is fed between the two parts, which are spaced so as to produce the maximum bran removal coupled with the minimum kernel breakage. The product from this machine is *unpolished milled rice*: the outer bran layers have been removed, but not the inner layers. Alternatively, the bran can be removed by a wet-brushing process, as described for oat kernels (p. 216).

The unpolished milled rice is next polished in a brush machine which removes the aleurone layer and any adhering particles, and yields *polished rice*. The polished rice may be further treated by coating with sugar syrup and talc powder and tumbling in a machine called a Rice Trumbol in order to produce a brighter shine on the grains.

Rice that has been milled by a scouring process is liable to develop oxidative rancidity; the risk is minimized by the polishing process.

The broken grains are removed as by-products; the largest fragments are known as *second head*, the medium as *screenings*, the smallest as *brewers' milled rice*. The last is so called because the small broken rice kernels are used as an adjunct in brewing in order to improve shelf life of the beer, and to impart stability (cf. p. 204).

The yield of products obtained from the milling of paddy rice in the U.S.A. and in Burma is shown in Table 54.

TABLE 54

MILLING YIELDS FROM PADDY RICE*

Product	U.S.A. (average) (%)	Burma (parboiled) (%)
Head rice	57 ⎫	71·6 ⎫
First head	3·5 ⎬ 68·5	0·5 ⎬ 72·1
Second head	6	
Brewers' rice	2 ⎭	⎭

* Data from Grist (1959).

The milling equivalents (yields of milled rice) (cf. p. 1) quoted in *Grain Crops* (Commonwealth Economic Committee) range from 73% in Japan and Korea to 63–64% in Vietnam, Cambodia, Laos, the Philippines and Malaya. The milling equivalent for the U.S.A. is given as 69–70%.

Ordinary and glutinous rice

Ordinary rice has vitreous endosperm; another type of rice, known as glutinous, sweet or waxy rice, has chalky opaque endosperm, the cut surface of which has the appearance of paraffin wax. The starches in these two types of rice have differing characteristics: that from ordinary rice gives a blue colour with iodine, whereas the starch from glutinous rice gives a reddish-brown colour, and somewhat resembles starch from root crops. Glutinous rice starch contains less than 1% of amylose but instead has molecules of low or medium molecular weight which are profusely branched and intermediate in structure between amylopectin and glycogen (Mickus, 1959; cf. p. 40).

Cooking and consumption of rice

The peoples of different countries have varying preferences for types of rice: round-grain rice is preferred in Japan, Korea and Puerto Rico, possibly because the cooked grains are more adherent, whereas long-grain rice (e.g. Patna rice) is preferred in the U.S.A. People in most countries prefer white rice, but in India the preference is for red, purple or blue strains.

Many rice-consuming families are without baking facilities and prepare rice by boiling it. When water is added to milled rice some of the nutrients contained in the grain dissolve in the water. Thus, washing rice before cooking, use of excess water in boiling, and pouring off and discarding excess boiling water all result in a loss of nutrients. Rice is ideally cooked by boiling in a minimum of water so that all the water is absorbed by the rice, thus preventing any loss of nutrients.

Polished rice is better digested than unpolished, but parboiled rice is digested better still.

Nutritive value

Specimen figures for the chemical composition of rice and its milled products were given in Table 10 (p. 37).

Parboiling. The parboiling of rice is an ancient tradition in India, and consists of steeping the rough rice in hot water, steaming it, and then drying down to a suitable moisture content for milling. The original purpose of the process was to loosen the hulls, but in addition the nutritive value of the milled rice is increased by this treatment, because the water dissolves vitamins and minerals present in the hulls and bran coat and carries them into the endosperm. Thus, valuable nutrients which would otherwise be lost with the hulls and bran in rice milling are retained by the endosperm.

By gelatinizing the starch of the outer layers, parboiling seals the aleurone layer and scutellum, so that these portions of the grain are retained in milling to a greater degree in pɘɯ par-boiled than in milled raw rice. Parboiling toughens the grain and reduces the amount of breakage in milling. Moreover, parboiled rice is less liable to insect attack, and keeps better than milled raw rice.

Conversion

Conversion of rice is the modern development of parboiling.

In the H.R. Conversion Process, wet paddy is held for about 10 min in a large vessel which is evacuated to about 25 in. The paddy is then steeped for 2–3 hr in water at 75–85°C introduced under a pressure of 80–100 lb/in². The steeping water is drained off, and the paddy is heated under reduced pressure for a short time with live steam in a steam-jacketed vessel. The steam is blown off, and the pressure in the vessel reduced to 28–29 in. of vacuum. The product is thus vacuum dried to about 15% m.c. in the steam-jacketed vessel, or it can be air-dried at temperatures not exceeding 145°C. After cooling, the converted paddy is tempered in bins for 8 hr or more to permit equilibration of moisture, and is then milled.

In the Malek Process, paddy is soaked in water at 100°F for 4–6 hr, steamed at 15 lb/in² pressure for 15 min, dried and milled. The product is called Malekized rice.

Parboiled or converted rice is readily consumed in India, while in the U.S.A. it is used as a ready-to-eat cereal, as canned rice, and as a soup ingredient. However, parboiled rice is not popular elsewhere.

Figures for the vitamin content of rice bran and hulls before and after conversion are shown in Table 55.

TABLE 55

VITAMIN CONTENTS OF RICE HULLS AND BRAN BEFORE AND AFTER CONVERSION* (μg/g)

	Vitamin B$_1$		Riboflavin		Niacin	
	Hulls	Bran	Hulls	Bran	Hulls	Bran
Before	3·0	20·5	0·64	2·6	46·8	229
After	1·5	7·5	0·51	1·5	31·4	200

* Source: Kik and van Landingham (1943).

The vitamin contents of milled (unconverted) and parboiled (converted) rice and of some of the by-products of the milling of unconverted rice are shown in Table 56.

The milling and polishing of raw rice results in losses of 76% of the vitamin B$_1$, 56% of the riboflavin and 63% of the niacin, whereas after parboiling, these losses are reduced to 58, 34 and 11%, respectively. The folic acid content of rice is increased from 0·04 to 0·08 μg/g by conversion whilst that of the polishings is reduced from 0·26–0·40 to 0·12 μg/g (De Caro *et al.*, 1949).

Fortified rice. The nutritive value of milled non-parboiled rice is so low as to cause concern in countries, e.g. Puerto Rico, in which rice is the principal foodstuff. Both brown rice (with intact bran) and milled parboiled rice have higher nutritive values, but neither of these forms of rice is favoured by Puerto

TABLE 56

VITAMIN CONTENTS OF RICE AND RICE PRODUCTS*

(μg/g)

Material	Vitamin B$_1$	Riboflavin	Niacin
Paddy rice	3·5–4·0	0·4–0·5	52·3–55·0
Milled, polished (non-parboiled) rice	0·4–0·8	0·15–0·3	14·0–25·0
Converted, parboiled rice	1·9–3·1	0·3–0·4	31·2–47·8
From milled, non-parboiled rice:			
Rice hulls	0·8–1·3	0·62–0·95	45·0–48·5
Rice bran:			
1st break	21·0–33·0	2·0–3·3 ⎫	
2nd break	12·0–24·0	1·37–2·2 ⎬	201·0–258·0
Rice polishings	15·0–28·0	1·14–1·87 ⎭	

* Sources: Kik (1943). Kik and van Landingham (1943).

Ricans. For these people, rice is prepared by blending 1 pt of concentrated fortified white rice with 20 pt of commercial white (milled) rice. The white rice is fortified by moistening it with a solution of the required vitamins in acid solution and then coating the grains with a film of zein and abietic acid in alcoholic solution. After the film has dried the grains are dusted with ferric pyrophosphate and talc powder. The film prevents loss of the vitamins when the rice is cooked by traditional methods.

The minimum contents of nutrients specified in 1952 by the U.S. Government in brown, parboiled and fortified rice are shown in Table 57.

TABLE 57

U.S. MINIMUM NUTRIENT CONTENTS IN RICE (1952)

Product	Thiamin (vitamin B$_1$) (mg/lb)	Ribo-flavin (mg/lb)	Nicotinic acid (niacin) (mg/lb)	Iron (mg/lb)
Brown rice	1·35	0·3	16	13
Parboiled rice	1·2	0·15	16	13
Fortified rice	2·0	—	16	13

N.B. mg/lb $\times$ 2·2 = μg/g.

Riboflavin is not added to fortified rice because it causes discoloration.

By-products of rice milling

The average yields of by-products obtained in the U.S.A. from 100 pt of paddy rice are 20 pt of hulls, 8·5 pt of rice bran and 2 pt of polishings (Grist, 1959). Where the rice milling industry is well established these products are obtained separately, but in more primitive milling processes the "rice bran" may be grossly contaminated with hulls and even with low grade broken rice (brewers' rice). The inclusion of hulls in rice bran lowers its feeding value because of the high fibre and silica contents of the hulls and their low digestibility (cf. Table 21). The Association of American Food Control Officials has adopted the following standards for rice bran: not less than 11% of crude protein, not less than 10% of crude fat, and not more than 15% of fibre. The maximum content of rice hulls in rice "bran", in order to attain this standard, is about 20%.

Specimen figures for the chemical composition of the by-products of rice milling are shown in Table 58.

TABLE 58

CHEMICAL COMPOSITION OF MILLING BY-PRODUCTS FROM JAPONICA-TYPE RICE*

Material	Moisture (%)	Protein (%)	Fat (%)	Ash (%)	Fibre (%)	Carbo-hydrate (%)
Husk	6·1	2·7	0·9	20·1	36·1	34·1
Meal (bran)	9·4	12·8	15·1	11·3	13·5	41·8
Polish	8·7	11·4	8·8	5·3	2·0	63·7

*Source: Fraps (1916).

Rice bran oil. Rice bran has a high oil content (10–15%). When the bran is separated from the kernel, hydrolytic action by the enzyme lipase begins to split the fat to FFA and glycerol (cf. p. 213), and oxidation of the fat by peroxidases may also occur. Peroxides are undesirable in feedstuffs because they destroy fat-soluble vitamins and also cause digestive disturbances.

The Rice Growers' Association of California have established plants at Sacramento and at Houston, Texas, for the extraction of oil from rice bran. The oil is of higher value than the bran, and is used for salad oil and cooking oil. When refined, bleached and deodorized it has twice the stability of comparable commonly used vegetable fats.

The nutritive value of the extracted bran is, moreover, enhanced because the protein content is somewhat increased.

The composition of rice bran oil is shown in Table 20.

Rice hulls. The ash content of rice hulls is about 22%, of which 95% is silica, and most of the remainder lime and potash (cf. Table 21). Rice hulls find uses dependent on their abrasiveness (cf. p. 225) and their high insulating value. They are also used in the commercial manufacture of furfural, although their pentosan content is lower than that of oat hulls or maize cobs (cf. p. 225).

MAIZE

Utilization

Maize (*Zea mays*) is used for animal feeding, for human consumption, and for the manufacture of starch, syrup and sugar, industrial spirit and whisky (cf. p. 204). The products of milling include maize grits, meal, flour (corn flour), germ and germ oil, hominy feed, starch (and its hydrolysis products), protein (gluten feed) and corn steep liquor. The ready-to-eat breakfast cereal "corn flakes" is made from maize grits (cf. Ch. 11).

The U.S.A. is the biggest domestic user of maize. In 1962–3, out of a total supply of 132·1 million tons (consisting of 91·1 million tons of new crop plus 41·0 million tons of "carry over" or initial stocks), 9·8 million tons were exported and 89·9 million tons used domestically (leaving 32·4 million tons surplus as "carry over" to 1963–4). Of the maize consumed domestically, 94% was used for animal feed, dry milling and farm household use, 5% for wet milling to produce starch, sugar and syrup, 0·8% for alcohol, and 0·3% for seed.

Utilization of imported maize in the U.K. is also chiefly for animal feed (79% of the total in 1962–3). Other uses in that season were for starch and glucose (10%), brewing and distilling (8%), and breakfast cereals (3%) (data from *Grain Crops*).

Maize milling

Maize is milled by dry or wet processes. The first objective in both processes is separation of the germ from the remainder of the grain in order to extract and recover the germ oil; the oil is a valuable product when separated, but it could lead to the development of rancidity if allowed to remain as a constituent of maize meal.

After degermination, the dry milling employs rollermills and plansifters in a process somewhat resembling wheat milling, whereas the wet milling involves a steeping stage and the complete disintegration of the endosperm tissue in order to recover starch and protein as separate products.

Dry milling

The products of dry maize milling, their particle size ranges, and average yields are shown in Table 59.

TABLE 59

YIELD AND PARTICLE SIZE RANGE OF MILLED MAIZE PRODUCTS*

Product	Particle size range		Yield % by wt.
	Mesh	in.	
Grits	14–28	0·054–0·028	40
Coarse meal	28–50	0·028–0·0145	20
Fine meal	50–75	0·0145–0·0095	10
Flour	thro' 75	below 0·0095	5
Germ	3–30	0·292–0·0268	14
Hominy feed			11

* Source: Stiver, Jr. (1955).

Maize grits are approximately equivalent to coarse semolina in particle size.

The objectives in dry maize milling are to obtain the maximum yield of grits with the least possible contamination with fat and black specks of tip cap (cf. p. 28); to recover as much as possible of the remainder of the endosperm as meal, while making the minimum amount of flour; and to recover the maximum amount of germ in the form of large particles with the maximum oil content.

The sequence of operations in dry maize milling is as follows:

Cleaning
Conditioning
Degerminating
Drying and cooling
Grading and aspirating
Grinding on fluted rolls
Sifting and classifying
Purifying and aspirating
Drying
Packaging

Cleaning. The impurities found in maize are similar to those encountered in other cereals, and the methods adopted for cleaning maize resemble those described for wheat cleaning (cf. Ch. 6). A normal cleaning flow would include magnet, milling separator, scourer-aspirator, dry stoner or specific gravity table, electrostatic separator, washer with overflow arrangement for floating off light impurities, and whizzer.

Particularly serious contamination problems of maize relate to rodent excreta pellets and insect fragments. Maize grains are larger than the grains of other cereals, and overlap in size range with rat excreta pellets. Removal of mouse pellets, which are much smaller than maize grains, is less of a problem with maize than with other cereals, but removal of rat pellets is correspondingly more difficult. Stiver has quoted the efficiencies of various machines for the removal of rat pellets, and the amount of maize rejected as screenings at the same time (Table 60).

Sequential cleaning of the maize on various machines does not necessarily give a better result, because the same pellets, viz. those lightest and smallest, would probably be extracted by any of the various machines. The best solution to the problem of

TABLE 60

EFFICIENCY OF MAIZE CLEANING MACHINERY FOR REMOVAL OF RAT EXCRETA PELLETS*

Machine	Rat pellets removed†	Maize extraction‡
Milling separator	10–30	2
Length indent separator	30–50	2
Width indent separator	20–40	2
Scourer-aspirator	20	1
Specific gravity table	50–70	1
Air separator	50–70	1
Wet flotation	50–70	0·5
Electrostatic separator	90–100	0·5

* Source: Stiver, Jr. (1955).

† As percentage of those in the original uncleaned maize.

‡ As percentage of the feed to the machine.

rodent contamination is the storage of grain under rodent-proof conditions (cf. p. 130).

Conditioning. The objectives in conditioning maize for degerminating are to loosen and toughen the germ and bran, and to bring the endosperm to an ideal moisture content such that the yield of grits is at a maximum, and that of the flour at a minimum, in the subsequent milling. The conditioning process is similar to that used for wheat, viz., addition of cold or hot water, or steam, but the moisture content is raised to a higher level, viz. 20–22%. The grain then rests for 1–2 hr.

Degerminating. This process, really degerminating and dehulling, is carried out in most U.S. mills by means of a Beall degerminator. This consists of a cast iron cone rotating at about 750 rev/min within a conical stationary housing. The maize is fed in at the small end and works along to the large end, between the two elements. Protrusions on the rotor rub off the hull and germ by abrasive action, and break the endosperm into two or more pieces per grain.

Other devices used for degerminating are entoleters, and rollermills using rolls fluted 6–8 cuts/in. and a differential of $1\frac{1}{4}$:1 or $1\frac{1}{2}$:1. The entoleter (cf. p. 141), in comparison with the

Beall mill, is said to cause less fragmentation of the germ, but to make grits of smaller size; the rollermill makes a large proportion of small germ fragments.

Drying, cooling, grading. The stock from the degerminators, at 21–22% m.c., is dried to 15–15·5% m.c. in rotary steam tubes at a temperature of 140–160°F, cooled to 90–100°F by aspiration with cold air, and then sifted to produce a number of fractions, viz. large, medium and fine hominy, germ roll stock, and meal (in order of decreasing particle size). These stocks are then fed to the mill, each entering at an appropriate point.

Milling. The milling is carried out on rollermills, of which a typical flow might contain as many as 16 distinct stages, all of which use fluted rolls. The products are sifted on plansifters. The mill is divided into a break section, a series of germ rolls, and a series of reduction and quality rolls. The break system releases the rest of the germ as intact particles, and cracks the larger grits to produce grits of medium size. The whole milling system for maize bears some resemblance to the earlier part of the wheat milling system as far as B_2 reduction roll (2nd quality roll, in the U.S.A.) (cf. pp. 118–126), viz. the breaks, the coarse reduction and the scratch, but is extended and modified in comparison with this part of the wheat milling system to make a more thorough separation of the large quantity of germ present.

The large and medium hominy (3–5 wire and 5–8 wire particle size ranges, respectively) enter the mill at the first break, and the coarse scalpings from this and from each successive break go to the germ rolls. The fine hominy and the second germ roll stock (8–14 wire and 14–28 wire, respectively) go directly to the second germ roll: the scalpings from each germ roll go to the finished germ aspirator and thence to the product bag as "germ".

The second cut from each break goes to the next break, while the germ roll second cuts go to the next germ roll. The third cuts (throughs of 14 wire) are table grit size stock, and go, via an aspirator, to the drier.

The fourth cuts (throughs of 28 wire) are coarse meal and go to the quality coarse purifier. The fifth cut (throughs of 50 wire) is further sifted over 75 wire; the tailings are fine meal, the throughs flour.

All the finished grits, meal and flour products are dried to 12–14% m.c. on rotary steam tube driers.

The action of the rolls should be less severe at the head end of the mill than at the tail in order to cause as little damage as possible to the germ, so that the maximum yields of oil and oil-free grits may be obtained simultaneously.

Nutritive value

Maize has a lower nutritive value than wheat, in particular being deficient in the vitamin niacin (nicotinic acid)—cf. Table 23—and having a relatively low content of protein, which is deficient in lysine (like wheat) and in tryptophan (cf. p. 48). The disease pellagra, due to deficiency of niacin or niacinamide, is prevalent among peoples who rely on maize for a large proportion of their daily food. The distribution of niacin among the parts of the maize grain is shown in Table 25. In order to make maize products satisfactory as the main item of a dietary, the addition of lysine, tryptophan, vitamin B_1, niacin and riboflavin would be desirable.

The chemical composition of the products of maize milling is shown in Table 61.

The protein content of dry-milled maize flour is about 7%, whereas that of the wet-milled flour ("corn flour") is only about 0·7%. The wet-milled flour contains 97–98% of starch.

Oil extraction from germ. Oil is extracted either by mechanical pressing, e.g. with a screw press, or by solvent extraction. The germ from the mill is first dried to about 3% m.c. and then extracted while at a temperature of about 250°F. The extracted oil is purified by filtering through cloth, using a pressure of 80–100 lb/in². The oil, which is rich in essential fatty acids (cf. Table 20), has a sp. gr. of 0·922–0·925, and finds use as a

TABLE 61

CHEMICAL COMPOSITION OF MAIZE AND OF MILLED MAIZE PRODUCTS

Material	Yield (%)	Mois-ture (%)	Pro-tein (%)	Fat (%)	Ash (%)	Fibre (%)	Carbo-hydrate (%)	Liter-ature source*
Maize grain	100	10·8	10·0	4·3	1·5	1·7	71·7	1
Products from dry milling:								
Grits	40	14	9	0·8	—	—	—	2
Meal	30	12	8·9	4·9	1·0	1·2	72	1
Flour	5	12·6	7·1	1·3	0·6	0·9	77·5	1
Germ meal	14	10·8	13·0	12·5	3·6	4·1	56·0	3
Hominy feed	11	11·0	9·4	0·7	0·3	0·4	78·2	1
Products from wet milling:								
Corn flour	—	12·7	0·7	0·06	0·1	—	86·4	4
Gluten meal	—	9·1	35·5	4·7	1·1	2·1	47·5	3
Gluten feed	—	10·4	23·5	3·4	2·5	3·5	56·7	3

* 1. Woods (1907). 2. Stiver, Jr. (1955). 3. Woodman (1957). 4. Boundy (1957).

salad oil. Its high smoke point also makes it suitable for use as cooking oil.

Wet milling of maize

Maize is wet milled to obtain starch, oil, cattle feed (gluten feed, gluten meal, germ cake), and the hydrolysis products of starch, viz. liquid and solid glucose and syrup.

Operations. The sequence of operations in wet milling is shown in Fig. 40. Wet milling differs fundamentally from dry milling in being a maceration process in which physical and chemical changes occur in the nature of the basic constituents of the endosperm—starch, protein and cell wall material—in order to bring about a complete dissociation of the endosperm cell contents with the release of the starch granules from the protein network in which they are enclosed. In dry milling, the endosperm is merely fragmented into cells or cell fragments with no deliberate separation of starch from protein (except in protein displacement milling, which is a special extension of dry milling—· cf. p. 142).

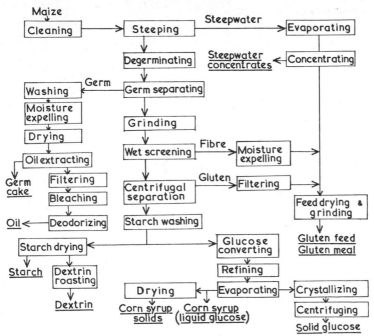

Fig. 40. Maize wet-milling process. (Adapted from Anon., *Food* **27**: 291, 1958; *Corn in Industry*, 5th edition, Corn Industries Research Foundation Inc., New York; and S. A. Matz (Ed.), *The Chemistry and Technology of Cereals as Food and Feed*, Avi Publ. Co., Westport, Conn., U.S.A. (1959).

Drying of maize. For safe storage, maize must be dried because the moisture content at harvest is generally higher than the desirable moisture content for storage. Grain for wet milling, however, should be dried at a temperature not exceeding 130°F; at higher temperature, changes occur in the protein whereby it swells less during steeping, and tends to hold the starch more tenaciously, than in grain not dried, or dried at lower temperature. In addition, if dried above 130°F, the germ becomes rubbery and tends to sink in the ground maize slurry (whereas the process of germ separation depends on the floating of the germ), and the starch tends to retain a high oil content.

Steeping. The cleaned maize is steeped for about 48 hr in warm water containing 0·02–0·03% of sulphur dioxide. The optimum temperature is about 50°C. Steeping is generally carried out in a number of wooden steeps, through which the steep water is pumped counter-current. The purpose of steeping is to soften the kernel and thereby facilitate separation of the hull, germ, fibre, protein and starch from each other. The moisture content of the grain increases rapidly to 35–40%, and more slowly to 43–45%.

Degerminating. After steeping, the steep water is drained off, and the maize coarsely ground in degerminating mills with the object of freeing the germ from the remainder of the grain without breaking or crushing the germ. The machine generally used for this purpose is a Fuss mill, a bronze-lined chamber housing two upright metal plates studded with metal teeth. One plate rotates at 900 rev/min, the other is stationary. Water and maize are fed into the machine, which cracks open the grain and releases the germ. By addition of starch–water suspension, the density of the ground material is adjusted to 8–10·5 Bé, which causes the germs to float while the grits and hulls sink.

Germ separation. The ground material flows down separating troughs in which the hulls and grits settle, while the germ overflows. The germ is washed and freed of starch on reels, de-watered in squeeze presses and dried on rotary steam driers. The dry germ is cooked by steam, and the oil extracted by hydraulic presses or by solvent. The germ oil is screened, filtered, and stored. The extracted germ cake is used for cattle feed.

The degerminated underflow from the germ separator is strained off from the liquor and finely ground on buhr stone mills, or on stainless steel vertical mills, such as the Bauer attrition mill. After this process, the starch and protein of the endosperm are in a very finely divided state and remain in suspension. The hulls and fibre, which are not reduced so much in size, can then be separated from the protein and starch by sieving. Fine fibres, which interfere with the subsequent separation of starch from protein, are removed on fine silk shakers.

Separation of starch from protein. In the raw grain, the starch is embedded in a protein network which swells during steeping and tends to form tiny globules of hydrated protein (Radley, 1951, 1952). Dispersion of the protein, which frees the starch, is accelerated by the sulphur dioxide in the steep water; the effect of the sulphur dioxide, according to Cox *et al.* (1944), is due to its reducing, not to its acidic, property. The sulphur dioxide also has a sterilizing effect, preventing growth of micro-organisms in the steep.

The suspension of starch and protein from the wet screening, adjusted to a density of 6 Bé by de-watering over Grinco or string filters, was formerly subjected to a process known as "tabling", a kind of continuous sedimentation, in which the starch settles out as a sludge while the protein is carried away, still suspended in the supernatant liquid. This process is no longer practised; instead, the starch and gluten are separated in continuous centrifuges.

Finally, the starch is filtered and dried. The moisture content is reduced to 10–12% by kiln-drying, and further reduced by vacuum drying to 5–7% in the U.S.A., or to 1–2% in Britain.

The protein in the steep water is recovered by vacuum evaporation, allowed to settle out of the water in tanks, and dried as "gluten feed" for animal feeding. The water recovered is re-used as steep water or, after concentration, is used as a medium for the culture of organisms from which antibiotics are obtained.

Maize cobs

The maize cob (corn cob in the U.S.A.) is the central rachis (cf. p. 21) to which the grains are attached and which remains as agricultural waste after threshing. As approximately 10 lb of cobs (d.m.b.) are obtained from every bushel (56 lb) of maize shelled, the annual production of cobs in the U.S.A. alone is of the order of 15 million tons.

Cobs consist principally of cellulose 35%, pentosans 40%, and lignin 15%; they are used agriculturally and industrially. The utilization of maize cobs has been thoroughly investigated by the N.R.R.L. (at Peoria, Illinois) of the U.S. Dept. of Agriculture. A report from this laboratory by Clark and Lathrop (1953) lists the following uses:

Agricultural uses. These include litter for poultry and other animals; mulch and soil conditioner; animal and poultry feeds; and, when ground to flour, diluents and carriers for insecticides and pesticides.

Industrial uses. Those based on physical properties are found in vinegar manufacture; fur cleaning; burnishing, polishing and soft grit blasting (cf. p. 225); absorbents, driers and floor sweeping compounds; abrasives for soaps; manufacture of asphalt shingles and roofing, bricks, and ceramics; oilwell drilling; fillers for explosives, plastics, glues, adhesives, rubber compounds and tyres.

Industrial uses based on chemical properties include manufacture of furfural (cf. p. 225); manufacture of fermentable sugars, solvents and liquid fuels; production of charcoal, gas and other chemicals by destructive distillation; use as solid fuel (according to Porter and Wiebe, 1948, oven-dry cobs have a calorific value of about 8000 Btu/lb); and in the manufacture of pulp, paper and board.

Sorghum

Utilization in the U.S.A. About 90% of the crop is used for animal feed. The remainder is used: (i) In wet milling, to make starch and its derivatives, with edible oil and gluten feed as by-products. The starch is used in food products, adhesives and sizings. (ii) In dry milling, to make a low-protein flour which is used for adhesives and oilwell drilling muds. The residue is used for feed. (iii) In the fermentation industry, for brewing, distilling and the manufacture of industrial alcohol (Martin and MacMasters, 1951).

REFERENCES

ASHLEY, W. (1928), *The Bread of our Forefathers: An Enquiry in Economic History*, Clarendon Press, Oxford.

Rye, Rice, Maize   255

BARGER, G. (1931), *Ergot and Ergotism*, Gurney & Jackson, London.

BOUNDY (1957), quoted in KENT-JONES, D. W. and AMOS, A. J. (1957).

CLARK, T. F. and LATHROP, E. C. (1953), *Corncobs—Their Composition, Availability, Agricultural and Industrial Uses*, U.S. Dept. Agric., Agric. Res. Admin., Bur. Agric. Ind. Chem., AIC–177, revised April 1953.

COX, M. J., MACMASTERS, M. M. and HILBERT, G. E. (1944), Effect of the sulphurous acid steep in corn wet milling, *Cereal Chem.* 21: 447.

DE CARO, L., RINDI, G. and CASELLA, C. (1949), Contents in thiamine, folic acid, and biotin in an Italian converted rice, *Abst. Communs. 1st Intern. Congr. Biochem.* 31.

FRAPS, G. S. (1916), The composition of rice and its by-products, *Texas Agr. Exp. Sta. Bull.* 191: 5.

GRIST, D. H. (1959), *Rice*, 3rd edition, Longmans, London.

KIK, M. C. (1943), Thiamin in products of commercial rice milling, *Cereal Chem.* 20: 103.

KIK, M. C. and LANDINGHAM, F. B. VAN (1943), Riboflavin in products of commercial rice milling and thiamin and riboflavin in rice varieties. The influence of processing on the thiamin, riboflavin and niacin content of rice, *Cereal Chem.* 20: 563, 569.

MCCANCE, R. A., WIDDOWSON, E. M., MORAN, T., PRINGLE, W. J. S. and MACRAE, T. F. (1945), The chemical composition of wheat and rye and of flours derived therefrom, *Biochem. J.* 39: 213.

MARTIN, J. H. and MACMASTERS, M. M. (1951), Industrial uses for grain sorghum, *U.S. Dept. Agric., Yearbook Agric., 1951*, p. 349.

MELLANBY, E. (1930), A lecture on the relation of diet to health and disease: some recent investigations, *Brit. Med. J.* i: 677.

MICKUS, R. R. (1959), Rice (*Oryza sativa*), *Cereal Sci. Today* 4: 138.

NEUMANN, M. P., KALNING, H., SCHLEIMER, A. and WEINMANN, W. (1913), Die chemische Zusammensetzung des Roggens und seiner Mahlprodukte, *Z. ges. Getreidew.* 5: 41.

PLANTE, E. C. and SUTHERLAND, K. L. (1944), Physical chemistry of flotation. X. Separation of ergot from rye, *J. Phys. Chem.* 48: 203.

PORTER, J. C. and WIEBE, R. (1948), *Gassification of Agricultural Residues*. U.S. Dept. Agric., Bur. Agric. Ind. Chem., North Reg. Res. Lab., Mimeo Circ. AIC–174.

RADLEY, J. A. (1951–2), The manufacture of maize starch, *Food Manuf.* 26: 429, 488; 27: 20.

STIVER, T. E., JR. (1955), American corn-milling systems for degermed products, *Bull. Assoc. Oper. Millers* 2168.

VAVILOV, N. (1926), Studies on the origin of cultivated plants, *Bull. Appl. Bot. Plant Breeding (Leningrad)* 16: 139.

WASSERMAN, T., FERREL, R. E., BROWN, A. H. and SMITH, G. S. (1957), Commercial drying of western rice, *Cereal Sci. Today* 2: 251.

WOODMAN, H. E. (1957), *Rations for Livestock*, Ministry of Agriculture, Fisheries and Food, Bull. 48, H.M.S.O., London.

WOODS, C. D. (1907), Food value of corn and corn products, *U.S. Dept. Agric. Farmers Bull.* 298.

FURTHER READING

ANON., A new maize grinding plant, *Food* **27**: 291, 1958.

CLARK, T. F. and ASHBROOK, J. W., Why the emphasis on corn cobs?, *Chemurg. Dig.*, June–July 1953.

MÖINICHEN, E., Rye for breadmaking, *Chem. & Ind.*, p. 496, 1953.

SCHOPMEYER, H. H., Rye and rye milling, *Bakers' Dig.*, p. 66, June 1959; *Cereal Sci. Today* **7**: 138, 1962.

VAHEY, L., Rye and rye milling, *Milling* **118**: 322, 1952.

WOLF, M. J., MacMASTERS, M. M., CANNON, J. A., ROSEWALL, E. C. and RIST, C. E., Preparation and properties of hemicelluloses from corn hulls, *Cereal Chem.* **30**: 451, 1953.

See also References to Chapter 3.

FURTHER READING—GENERAL

BACHARACH, A. L. and RENDLE, T. (Ed.), *The Nation's Food*, Soc. Chem. Ind., London, 1946.

BAILEY, C. H., *Constituents of Wheat and Wheat Products*, Reinhold Publ. Co., New York, 1944.

BATE-SMITH, E. C. and MORRIS, T. N. (Ed.), *Food Science*, Cambridge University Press, 1952, ch. 2 (part): Cereals (J. J. C. HINTON, W. J. S. PRINGLE, E. N. GREER and N. L. KENT).

HORDER, T., DODDS, E. C. and MORAN, T., *Bread*, Constable, London, 1954.

JACOBS, M. B., *The Chemistry and Technology of Food and Food Products*, Interscience Publ., New York, 1944.

KENT-JONES, D. W. and AMOS, A. J., *Modern Cereal Chemistry*, 5th edition, Northern Publ. Co. Ltd., Liverpool, 1957.

MATZ, S. A. (Ed.), *The Chemistry and Technology of Cereals as Food and Feed*, Avi Publ. Co., Westport, Conn., U.S.A., 1959.

Symposium on The quality of cereals and their industrial uses, *Chem. & Ind.*, pp. 31, 46, 75, 95, 112, 117, 144, 160, 185, 289, 389, 496, 578, 787, 890, 1953.

WINTON, A. L. and WINTON, K. B., *Structure and Composition of Foods:* Vol. 1, *Cereals, Nuts, Oil Seeds*, Wiley & Sons, New York, 1932.

INDEX